SpringerBriefs in Physics

Series Editors

Balasubramanian Ananthanarayan, Centre for High Energy Physics, Indian Institute of Science, Bangalore, India

Egor Babaev, Physics Department, University of Massachusetts Amherst, Amherst, MA, USA

Malcolm Bremer, H H Wills Physics Laboratory, University of Bristol, Bristol, UK

Xavier Calmet, Department of Physics and Astronomy, University of Sussex, Brighton, UK

Francesca Di Lodovico, Department of Physics, Queen Mary University of London, London, UK

Pablo D. Esquinazi, Institute for Experimental Physics II, University of Leipzig, Leipzig, Germany

Maarten Hoogerland, University of Auckland, Auckland, New Zealand

Eric Le Ru, School of Chemical and Physical Sciences, Victoria University of Wellington, Kelburn, Wellington, New Zealand

Dario Narducci, University of Milano-Bicocca, Milan, Italy

James Overduin, Towson University, Towson, MD, USA

Vesselin Petkov, Montreal, QC, Canada

Stefan Theisen, Max-Planck-Institut für Gravitationsphysik, Golm, Germany

Charles H.-T. Wang, Department of Physics, The University of Aberdeen, Aberdeen, UK

James D. Wells, Physics Department, University of Michigan, Ann Arbor, MI, USA

Andrew Whitaker, Department of Physics and Astronomy, Queen's University Belfast, Belfast, UK

SpringerBriefs in Physics are a series of slim high-quality publications encompassing the entire spectrum of physics. Manuscripts for SpringerBriefs in Physics will be evaluated by Springer and by members of the Editorial Board. Proposals and other communication should be sent to your Publishing Editors at Springer.

Featuring compact volumes of 50 to 125 pages (approximately 20,000–45,000 words), Briefs are shorter than a conventional book but longer than a journal article. Thus, Briefs serve as timely, concise tools for students, researchers, and professionals. Typical texts for publication might include:

- A snapshot review of the current state of a hot or emerging field
- A concise introduction to core concepts that students must understand in order to make independent contributions
- An extended research report giving more details and discussion than is possible in a conventional journal article
- A manual describing underlying principles and best practices for an experimental technique
- An essay exploring new ideas within physics, related philosophical issues, or broader topics such as science and society

Briefs allow authors to present their ideas and readers to absorb them with minimal time investment.

Briefs will be published as part of Springer's eBook collection, with millions of users worldwide. In addition, they will be available, just like other books, for individual print and electronic purchase.

Briefs are characterized by fast, global electronic dissemination, straightforward publishing agreements, easy-to-use manuscript preparation and formatting guidelines, and expedited production schedules. We aim for publication 8–12 weeks after acceptance.

More information about this series at http://www.springer.com/series/8902

Oleg V. Minin • Igor V. Minin

The Photonic Hook

From Optics to Acoustics and Plasmonics

 Springer

Oleg V. Minin ⓘ
Tomsk Polytechnic University
Tomsk, Russia

Igor V. Minin ⓘ
Tomsk Polytechnic University
Tomsk, Russia

ISSN 2191-5423   ISSN 2191-5431   (electronic)
SpringerBriefs in Physics
ISBN 978-3-030-66944-7   ISBN 978-3-030-66945-4   (eBook)
https://doi.org/10.1007/978-3-030-66945-4

This Springer imprint is published by the registered company Springer Nature Switzerland AG
The registered company address is: Gewerbestrasse 11, 6330 Cham, Switzerland

Dedicated to our Father[1], friend and colleague, all in one,
and equally to our Mother with love and gratitude.
Without your help and support,
this book would never have been written.
You are the greatest factor contributing to our academic accomplishments.
Father would have been most proud.

---

[1] Prof. Vladilen F. Minin. https://en.wikipedia.org/wiki/Vladilen_F._Minin

Foreword

The idea that light propagates along straight lines was known since antiquity – an aspect explicitly stated in Euclid's "Optica" written around 300 BCE. More than four decades ago, the possibility of a freely accelerating wavepacket was first suggested within the context of quantum mechanics. This is possible even in the absence of any external potential as long as the wavepacket assumes the form of an Airy function. In 2007, this Airy self-acceleration process was first suggested and experimentally observed for the first time in optics. Ever since, this class of accelerating or self-bending beams has attracted considerable attention and found applications in many fields of optics.

Quite recently, a new family of near-field localized curved light beams has been discovered by the Minin brothers, which is fundamentally different from the family of Airy beams since they arise in the near-field zone. These beams are better known as "photonic hooks". In general, photonic hooks manifest themselves on a scale much smaller than that associated with Airy beams. Subsequently, many of the attributes related to photonic hook processes were successfully extended and experimentally observed in many other areas, like terahertz waves, surface plasmons and acoustic waves. In a photonic hook, the wavefront is shaped in such a fashion that the maximum energy transport in space follows a substantially curved trajectory. Photonic hooks are unique in the sense that their radius of curvature is considerably smaller than the wavelength, implying that such structured light beams can experience maximum acceleration. In other words, a photonic hook can be directly observed even in the shadow of a mesoscale particle and is free from any other limitations.

The realization of such curved near-field beams is possible because of the interaction of a plane wave with a mesoscale dielectric particle that happens to have a broken internal symmetry. Importantly, these new possibilities suggest that the concept of wave acceleration is universal since these effects can be observed both in near-field and far-field arrangements. On a more fundamental level, the introduction of the photonic hook concept into the fields of optics, acoustics and plasmonics will help the scientific community to better understand the physics of accelerating near-field waves and to manipulate light and acoustic wave interactions in unusual ways as a means to open up new paths for new applications.

This area is just beginning to emerge. I hope the reader will find this book to be an interesting introduction in structured curvilinear beam dynamics and benefit from their applications in optics, plasmonics and acoustics.

CREOL-The College of Optics and Photonics   Demetrios Christodoulides
University of Central Florida,
Orlando, FL, USA

Bending the Rules with Self-Bending Light

The twentieth century has come to be known as the "century of the electron", with electronic devices becoming the mainstay of many technological advances. Older generations will recall telephone networks as long copper wires and home television sets as large tube objects, both guiding electrons. Towards the end of the past century, advances in the manipulation and harnessing of light began to take hold, accelerating into the twenty-first century, now referred to as the "century of the photon", with electronic devices steadily replaced by photonic devices. We now communicate with photons down fibre optics, and our favourite movie is surely watched on a light-emitting-diode (LED), or liquid-crystal-display (LCD) television, both photonic devices.

But for all the marvellous advances, photonic devices still lag far behind their electronic counterparts, primarily because the toolkit for manipulation and control of photons is still far from developed. Early optical control made use of reflection and refraction, and later diffraction as computing power and fabrication advances made exotic phase structures in materials possible, now referred to as diffractive optical elements. Using interference and diffraction as a tool in the manipulation of light has heralded a step-change in manipulation. The power is to tailor the light from its intrinsic properties at the source to some desired set of properties, in principle in all degrees of freedom of the light. This topical research area is called structured light: sculpturing light much like a tailor might sculpture a gown out of cloth. Unlike cloth, light can be structured in 3+1 dimensions: all three spatial dimensions as well as in time. Control in the three special dimensions can be achieved by a variety of approaches, by perhaps the most sophisticated is to carefully tailor the initial structure to have some predetermined evolution due to interference. Bessel beams are the most common examples of this, constructed by the interference of plane waves travelling on a cone, while waves of arbitrary trajectory and geometry can in turn be constructed by interfering multiple Bessel beams.

An exciting prospect of structured light is to "bend the rules" of physics by tailoring the entire field but considering only a region of interest within it: Bessel beams appear non-diffracting and Airy beams appear to follow a curved path if we only consider the central intensity lobe and not the entire field. This offers a new route to exotic photonic behaviour while still obeying Maxwell's equations. In parallel are near-field effects, on the order of the wavelength of the light, where intuition often fails. One such example is the recent notion of photonic jets, when light passing through meso-scaled symmetric objects produces sub-wavelength scale focusing. Combining these notions produces what is now known as *photonic hooks*, light that appears to bend in free-space and have features much smaller than the wavelength, yet requiring only meso-scaled asymmetric structures, for example, a wedge appended to a cube. This novel self-bending wave was demonstrated experimentally only recently (2019) with the aid of THz technology, and has shown potential to tailor wave fields across the wavelength range and for a variety of applications, and is fast gaining traction in the community (as recently highlighted by Nature Photonics).

This timely book introduces this exotic structured wave field, from light to sound, and provides a useful resource for graduate students and researchers alike who wish to familiarize themselves with the basics of the field. It covers the physical principles and applications, the creation and control of such self-bending fields from acoustic to optical waves, touching on plasmonic realizations, and highlights the exciting applications, particularly in nanoparticle manipulation. It is written authoritatively by authors who founded the field, while bringing in perspectives from others who have entered it.

As structured light takes shape, and photonics takes centre stage in our technological world, books that highlight what is possible are both welcome and instructive. I am delighted to see this comprehensive monograph take shape and promote the field to the next generation of researchers looking for new avenues to explore. As we move towards integrated photonics, so near-field control will become ever more important. Photonic hooks may very well be the peg that binds the photonic and material worlds together.

School of Physics   Andrew Forbes
University of the Witwatersrand
Johannesburg, South Africa

Acknowledgements

We are greatly indebted to all our co-authors and friends who took part in the preparation of the relevant articles on the subject of the book. We would especially like to express our gratitude to Drs. Z. Wang and L. Yue (School of Computer Science and Electronic Engineering, Bangor University, UK), for their contribution to optical hook; Drs. C. Rubio and A. Uris (Centro de Tecnologías Físicas: Acústica, Materiales y Astrofísica, Universitat Politècnica de València, Spain), for their contribution to acoustojet and acoustic hook investigations; Dr. A. Karabchevsky (Ben-Gurion University of the Negev, Israel); and also to Drs. D. Ponomarev, K. Zaytsev and Prof. Y. Geintz (Russia), for help in simulations and discussions in our joint works. We are infinitely grateful to Prof. Demetrios N. Christodoulides (CREOL-The College of Optics and Photonics, University of Central Florida, USA) and Prof. Andrew Forbes (School of Physics, University of the Witwatersrand, South Africa), for their willingness to write the forewords to this book, and also to Sam Harrison at Springer, for constant attention and help in our work with the book.

This work was partially supported by the Russian Foundation for Basic Research (Grants No. 20-57-S52001, 21-57-10001)

Introduction

The notion of controlling light has a very long history. The biblical account of the formation of the universe begins as "Let there be light!" [1]. Writing circa 300 BCE, Euclid introduced the abstract idea of a light ray. He formulated the law of rectilinear propagation of light in his treatise "Optics" [2]. Roman and ancient Greek historians recorded that during the siege of Syracuse in 212 BCE, Archimedes constructed a burning glass to set fire to the Roman warships. According to ancient writings, using large mirrors, Archimedes focused the sun's rays on the ships and reduced them to cinders. Intrigued by the idea, in October 2005, MIT's students confirm this myth experimentally [3]. This was probably the first example of the tailoring of light and practical implementation of structured light beams. In this regard, from the common point of view, lenses (optical lenses are known since 750BCE–710BCE – the so called Nimrud/Layard Lens [4]) or focusing mirrors, developed by the ancient Egyptians and Mesopotamians, may be considered as simplest forms of tailoring light [5]. Huygens introduced the wave nature of light in 1690 [6]. Later, the development of Maxwell's electrodynamics [7] reinforced the notions that light propagates along straight lines by ensuring the conservation of electromagnetic momentum and spreads through diffraction – and they cannot go around corners.

However, the possibility that a wavepacket can freely accelerate even in the absence of an external force was first discussed within the context of quantum mechanics in 1979 [8] – this is only possible as long as the quantum wavefunction follows an Airy-function profile. In 2007, this Airy self-acceleration process was first suggested and experimentally observed for the first time in optics [9, 10]. In the past few years, other types of Airy-like accelerating curved beams, which are another example of structured field, have been intensely explored in optics, acoustics and plasmonics [11]. In all cases, these Airy-like wavefronts propagate on a ballistic trajectory over several Rayleigh lengths while defying diffraction effects. Until recently, they provided the only example of "curved light transport" in nature.

Nevertheless, Airy family beams can only be bent through relatively small angles [11]. This means that they cannot provide the sharp turns needed for manipulation on the micron or even nanometre scale. At the other extreme, in the early 2000s, the phenomenon of the photonic nanojet (PNJ) began to be intensively studied [12] – this effect involves the interaction of radiation with dielectric micro-object or material structures of intermediate scale (Mie size parameter $q = 2\pi r/\lambda \approx (3\text{--}20\pi,$ where λ is illuminating wavelength and r is particle radius) that are too large to be characterized as simple dipoles and too small to be described by geometrical optics. However, all photonic nanojets are fundamentally similar – straight and single-coloured. We have helped to break that mould in 2015, when UNESCO held the International Year of Light and Light-Based Technologies, by discovering that light can indeed be bent by interaction with a mesoscale Janus dielectric particle, and with diffraction much less than a "regular" beam – the first curved photonic nanojet known now as photonic hook (PH) was discussed in [13]. This first simple structured field was constructed from combinations of dielectric mesoscale cube diffraction with prism refraction [13], which offers an intriguing ability to form a localized curved beam near the shadow surface of dielectric particle. In contrast to classical photonic nanojet, the photonic hook may be considered as a new kind of PNJ produced by the Janus particles with asymmetric shape, asymmetric internal structure or a symmetric particle with asymmetric illumination – the subwavelength localization of light in curved space can be controlled and manipulated in unprecedented ways. Such symmetry breaking can be used for the further improvement of the optical properties of the localized beams.

A distinctive feature of the PH is [14]: the radius of curvature of PH is a fraction of the illumination wavelength, and although the curved profile evokes a similarity with Airy beams, in contrast to Airy beams there are no curved side lobes [15, 16], where in the caustic from one side the sidelobes are almost parallel to each other. Moreover, in the PH, there is an inflexion where the curved beam changes its propagating direction. This property is not possessed by the Airy-like beam [14]. It is important to note that the PH structured field combines the construction simplicity of the PNJ, as well as the curvature produced by Airy-family self-bending beams, but at subwavelength level. We also note that previously studied curved beams usually require the use of expensive and complicated optical elements, which often make them unsuitable for embedding in an optical system. The PH concept does not need to apply any metasurfaces or dielectric or/and metallic subwavelength structures and PHs do exhibit subwavelength field localization, making them possible to use in highly integrated optical circuits.

Importantly, the photonic hook phenomenon was further extended to the fields of optics, terahertz waves, surface plasmons and acoustic waves [17–21]. It is well known that an electromagnetic wave is a transverse wave that can be demonstrated through polarization. At the same time, sound is a longitudinal mechanical wave that is a variation in pressure. Surface plasmon polaritons are electromagnetic waves that travel along a metal–dielectric interface with a wavelength shorter than of illuminated light, that is, low-dimension 2D waves. And for all these different types of waves, a new effect of structured near-field hook-type fields has been demonstrated [22].

Tailoring the complex trajectories in this photonic hook beams leads to the discovery and design of new curved beam families, as well as intriguing electromagnetic phenomena, such as the combination of both quasi-nondiffracting behaviour and abruptly varying direction in a subwavelength localized beam. By controlling beam curvature by field polarization, illuminating wavelength or illuminated field structurization, the technique could accommodate different wave-scaled applications, including optomechanical manipulation of nanoparticles, particle acceleration and separation, terahertz generation, near-field microscopy and spectroscopy, laser machining of various guiding structures to include wavelength scaled division multiplexers, and beamsplitting and beam-coupling.

This work not only reveals the explicit physical role of any given types of photonic hooks in optics, plasmonics and acoustics, but to our point also provides an alternative design roadmap of subwavelength curved structured light. This area of photonics, optics and acoustics is developing rapidly, and this book, which is truly based on the personal ideas and experiences of the authors, is just an introduction to it.

References

1. Genesis (1:3)
2. H. E. Burton. The Optics of Euclid. // Journal of the Optical Society of America, 35(5), 357–372 (1945)
3. web.mit.edu/2.009/www/experiments/deathray/10_ArchimedesResult.html
4. The Nimrud lens/The Layard lens. Collection database. The British Museum. http://www.britishmuseum.org/research/collection_online/collection_object_details.aspx?objectId=369215&partId=1
5. A. Forbes. Structured Light from Lasers. // Laser Photonics Rev. 1900140 (2019)
6. C. Huygens, *Traité de la Lumière*, Leiden: Pieter van der Aa, 1690
7. James Clerk-Maxwell. *A Treatise on Electricity and Magnetism.* (Clarendon Press Series, Macmillan & Co., 1873.)
8. M.V. Berry, N.L. Balázs, Non-Spreading Wave Packets // American Journal of Physics **47**, 264 (1979).
9. G.A. Siviloglou, J. Broky, A. Dogariu, D.N. Christodoulides, Observation of Accelerating Airy Beams // Physical Review Letters 99, 213901 (2007).
10. G.A. Siviloglou, D.N. Christodoulides, Accelerating finite energy Airy beams // Optics Letters **32**, 979 (2007)
11. N.K. Efremidis, Z. Chen, M. Segev, D.N. Christodoulides, Airy beams and accelerating waves: an overview of recent advances // Optica, 6(5), 686–701 (2019)
12. B. Luk'yanchuk, R. Paniagua-Domínguez, I. V. Minin, O. V. Minin, and Z. Wang, Refractive index less than two: photonic nanojets yesterday, today and tomorrow (Invited), Optical Materials Express 7(6), 1820–1847 (2017).
13. I. V. Minin, O. V. Minin, *Diffractive Optics and Nanophotonics: Resolution Below the Diffraction Limit*, Springer, Cham (2016).
14. I. V. Minin and O. V. Minin. Recent Trends in Optical Manipulation Inspired by Mesoscale Photonics and Diffraction Optics // J of Biomedical Photonics & Eng 6(2), 020301 (2020)
15. L. Yue, O. V. Minin, Z. Wang, J. Monks, A. Salin, and I. V. Minin, Photonic hook: a new curved light beam // Optics Letters 43(4), 771–774 (2018).
16. K. Dholakia, G. Bruce, Optical hooks // Nature Photonics 13(4), 229–230 (2019).

17. I. V.Minin, O. V. Minin, G. Katyba, N. Chernomyrdin, V. Kurlov, K. I. Zaytsev, L. Yue, Z. Wang, and D. N. Christodoulides. Experimental observation of a photonic hook // Appl. Phys. Lett. 114, 031105 (2019)

18. O. V. Minin, I. V. Minin, K. I. Zaytsev, G. Katyba, V. Kurlov, L. Yue, Z. Wang, "Electromagnetic field localization behind a mesoscale dielectric particle with a broken symmetry: a photonic hook phenomenon," Proc. SPIE 11368, Photonics and Plasmonics at the Mesoscale, 1136807 (2 April 2020)

19. I.V. Minin, O.V. Minin, D.S. Ponomarev, I.A. Glinskiy, Photonic Hook Plasmons: A New Curved Surface Wave // Ann. Physik. 1800359 (2018)

20. I. V. Minin, O. V. Minin, I. Glinskiy, R. Khabibullin, R. Malureanu, D. Yakubovsky, V. Volkov, D. Ponomarev. Experimental verification of a plasmonic hook in a dielectric Janus particle // arXiv:2004.10749 (2020)

21. C. Rubio, D. Tarrazó-Serrano, O. V. Minin, A. Uris, I. V. Minin. Acoustical hooks: A new sub-wavelength self-bending beam // Results in Physics 16, 102921 (2020)

22. I.V. Minin, O.V. Minin. Dielectric particle-based strategy to design a new self-bending sub-wavelength structured light beams.//IOP Conf. Ser.: Mater. Sci. Eng. 1019, 012093 (2021)

Contents

Chapter 1
Photonic Hook Main Properties

Abstract The study of accelerating Airy-family beams has made significant progress, not only in terms of numerical and experimental investigations but also in
conjunction with many potential applications. However, the curvature of such
beams (and hence their acceleration) is usually greater than the wavelength.
Relatively recently, a new type of localized wave beam with subwavelength curvature, called the photonic hook (PH), was discovered. This photonic hook is a curved
high-intensity focus by, for example, a wavelength-scaled dielectric cuboid topped
with a wedge prism (Janus particle) illuminated by a plane wave – an interesting
situation combines diffraction with prism refraction, which adds a newfound degree
of simplicity. The difference between the phase velocity and the interference of the
waves inside the particle causes the phenomenon of focus bending. It is important
that in the PH, there is an inflexion where the curved beam changes its propagating
direction. This property is not possessed by Airy-like beams family. Amazingly, the
mesoscale dielectric Janus particle, with broken shape or refractive index symmetry
or broken symmetry of illuminating wave, is used to generate the photonic hook –
emerging from its shadow-side surface. Moreover – PH has unique features – the
radius of curvature is less than the wavelength; this is the smallest curvature radius
of electromagnetic waves ever reported today. Although the curved profile evokes a
similarity with Airy beams, in contrast to Airy beams there are no curved sidelobes.
In this chapter, we discussed the key properties of PH based on full-wave simulations confirmed by experimental research of a scale model. The principal possibility
of exotic looped shape of localized field and photonic trajectories is also briefly
discussed. Using the photonic hook, the resolution of optical scanning systems can
be improved to develop optomechanical tweezers for moving nanoparticles, cells,
bacteria, and viruses along curved paths and around transparent obstacles. These
unique properties of photonic jets and hooks combine to afford important applications for low-loss waveguiding, subdiffraction-resolution nanopatterning, and
nanolithography.

1.1   Introduction

A common optical scheme for Airy beam formation includes a cylindrical lens and a spatial light modulator (SLM), which operates only at restricted power. The latter is mounted in the front focal plane of the cylindrical lens, while the Airy beam generates in its back focal plane and nearby. The cubic phase modulation of an incident Gaussian beam generated by a modulator acts as a cubic lens, whereas the curved path is determined by the acceleration rate of the cubic phase change [1]. The Airy laser beams are generated also by binary phase diffractive optical elements [2] and by the phase diffraction gratings, which are used as SLM analogs and create bending diffraction orders [3, 4]. The metasurfaces are also used for generating the Airy beams; however, the manufacturing process of the metasurfaces is rather complicated and expensive [5].

It is important that the diameter of the Airy beams is usually several incident wavelengths, equaling the diameter of an optical element which is much larger than the wavelength [2]. The curvature of the Airy-like beams is determined by a parameter that determines the rate of rise of the cubic phase [2] and is usually greater than several wavelengths. Scaling the Airy beam generation from visible light to the terahertz range is not always possible. This is because SLMs cannot operate in the terahertz range, due to the absence of materials with the required modulation [6]. Moreover, the main lobe of a finite energy Airy beam is not observed directly behind the cubic phase element and transition region, where the initial intensity distribution of the incoming beam is transformed into the distinct Airy pattern [7]. The complex classical Airy-like beam profile is arranged with a series of sidelobes or spikes, propagating along a parabolic trajectory.

Below we consider the mesoscale particles for the generation of photonic jets. The Mie parameter ($q = 2\pi r/\lambda$) is used to determine the particle size. Between the nano-scale optics ($q \sim 1$) and traditional optics ($q > 100$), there is an intermediate range ($q \sim 10\ldots40$) of the particle size in particular. In the visible light region, this intermediate range matches the micron-sized particles (from several micrometers to tens of micrometers). The photonic nanojets (PNJs) are earlier detected by using the particles in this intermediate range [8–10]. The generation of bounded light beams in the near-field is crucial for many applications, for example, for optical memory systems.

1.2   Near-Field Curved Beams

Following Ref. [11], let's briefly consider the diffraction on the dielectric edge. It is well known that when diffraction arises from the dielectric edge of the Fresnel zone plate, the edge wave has the eikonal approximation ($S \sim Z + X^2/2Z$), where X is the transverse coordinate and Z is the longitudinal coordinate [12]. A solution of the diffraction at the edge of the semi-infinite opaque screen is a paraxial two-dimensional light, for which the argument for the complex amplitude function depends on variables (X^2/Z) [13]. Instead of the Airy beams, the optical beam has a parabolic path $y = x^2$, whereas the considered edge beams in the equation propagate along a root-parabolic path $y = \sqrt{x}$. Figure 1.1 shows the full-wave simulation results of the diffraction on a rectangular phase plate, under the illumination of a linearly polarized plane wave. It can be seen in Fig. 1.1a that the curvilinear field localization areas arise in the region of the plate edge. As the plate width decreases, these localization regions approach each other. In cases of a small plate width, these optical beams begin to interfere with each other and form a central classical photonic jet during diffraction via a phase step [14, 15]. It should be noted that such curved beams can have the self-healing property of the Airy beams, as shown in Fig.1.1d.

It is obvious that by decreasing the width of the phase step, the side jets will coincide, forming a single local maximum, called the photonic jet. In Fig. 1.2 the formation of the photonic jet by three-dimensional (3D) dielectric cuboid with different refractive index is shown.

One can see a cubic particle can achieve a photonic jet [14–23]. But it is not a classical particle lens; to form a photonic jet, the following phenomena are involved: refraction, diffraction (scattering), and interference. Moreover, the laws are different from ray tracing. Photonic jet also exhibits some unusual properties not found in cylindrical or spherical particles, such as the effects of anomalous apodization [20, 22] and the unusual properties [24, 25] of the Gouy phase [26]. In common, the radiation passing through a transparent dielectric object possesses a phase delay compared to the radiation passing around such an object. Therefore, the wave front becomes concave and convergent, so the focusing condition for a wave could be achieved [27]. It could be noted that mesoscale particle with refractive index near 2 (Fig. 1.2c) finds interesting applications for nanofocusing, nanoparticle manipulation, and radiation modulation [28–30].

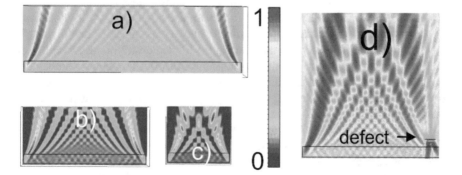

Fig. 1.1 Full-wave simulation results of the diffraction on a rectangular phase plate, with widths of (**a**) 36 λ, (**b**) 24 λ, (**c**) 12 λ, and (**d**) a rectangular phase plate with defect. Reprinted from [11] under the Creative Commons Attribution License

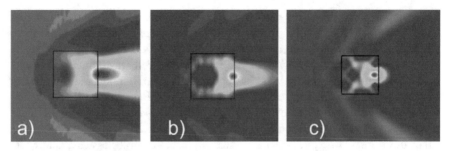

Fig. 1.2 Full-wave simulation results of the photonic jet formation by dielectric cuboid particle with different refractive index contrasts: 1.41 (**a**), 1.72 (**b**), 2.0 (**c**)

1.3   The Photonic Hook Phenomenon

Symmetry breaking of dielectric structure can be used for improvement of the optical properties and the shape of the localized field [27]. The key difference between the photonic hook (highly intensive wavelength-scaled curved light paths or curved power flow) and the Airy beam is that the photonic hook is created using the focusing of an electromagnetic wave through a Janus dielectric particle directly near the shadow surface of the particle, which in its simplest form is the combination of a wedge prism and a cuboid [27]. The influence of a prism angle and size of the Janus particle on the distribution of field enhancement, curvature of the photonic hook, length, and full width at half maximum (FWHM [31]) are effectively quantified for the first time in [32].

For simplicity and taking into account the results of [23], hereinafter we consider two-dimensional (2D) geometry (square rod with triangle prism) of the structure. To simulate it we used finite element method (Comsol Multiphysics software). The main simulation results are shown in Figs. 1.3 and 1.4. Figure 1.3 demonstrates the dynamics of the formation of the structure of a localized field in the shadow region of a combined particle, depending on the angle of the prism. Three characteristic areas can be distinguished. At small angles of the prism (less than 3.5°), the field structure coincides with the structure of the photonic jet. At angles of the prism from 4 to 20° (for the selected sizes of a square particle), a photonic hook region of various spatial shapes is formed. To this end Figure 1.4 demonstrates the curvatures of the PH vs prism angle. And at large angles (more than 30°), an oblique photonic jet is formed – the combinations of dielectric mesoscale square particle diffraction with prism refraction allow deflecting photonic jet in the case the prism angle is more than critical, which is clearly demonstrated in Fig. 1.3f.

The optical photonic hook formation by 3D cuboid particle with different dimensions topped with a wedge prism (triangular prism) is described in details in [32]. The Janus structure represents a dielectric cuboid with the addition of a wedge-shaped prism on the face encountering plane wave. The refractive indices of the particle material (transparent fused silica [33]) and background medium (air) were set up to 1.46 and 1.0, respectively. Simulations shown that the prism on the illumi-

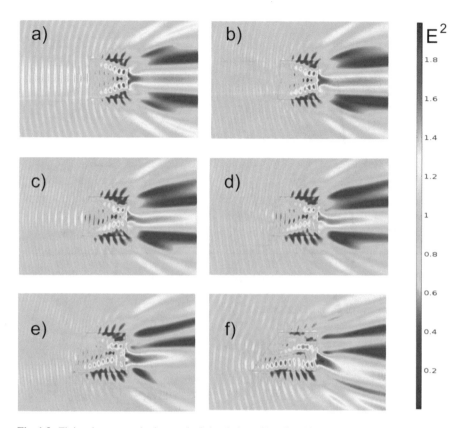

Fig. 1.3 Finite element method numerical simulation of localized field by square rod with triangle prism. The prism angles are 0° (**a**), 3.8° (**b**), 9.5° (**c**), 13° (**d**), 20° (**e**), 35° (**f**) degree

nated side of the cuboid not only lets the photonic jet shift away from the optical axis but also bends it in the central propagation, forming a specific "hook" shape. Moreover, the peak E^2 enhancement of the photonic hook is 1.25 times greater than that achieved by the photonic jet from a normal cuboid particle of the 3 wavelength dimensions. At the same time, the FWHM of the PH generated by the Janus particle is shorter than that for the PNJ induced by the cuboid particle [32] and is able to break the simplified diffraction limit (0.5λ criterion) [34–35].

Dependences of the curvature of the PH vs prism angle-α for 3D Janus particle based on wavelength-scaled cuboid shown the following [32]. According to [32], the curvature of the PH is defined by the factor-β and is the angle between the two lines linking the inflection point with the end point of the PH and the start point with the inflection point, respectively, as shown in Fig. 1.5. For example, for the cuboid dimensions of L = 3 of wavelength, when angle of a prism is increased from 0° to about of 5°, the symmetry of the PNJ becomes broken, and the photonic jet starts leaning to the direction of the longer side of the Janus particle, resulting in a 7° β-factor. The largest curvature, of about 35°, is achieved at a prism angle of about

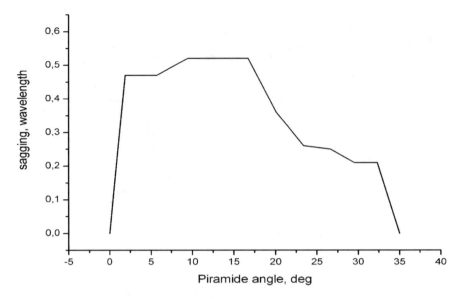

Fig. 1.4 Curvatures of the PH vs prism angle

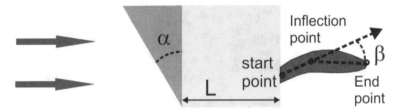

Fig. 1.5 Definition of the curvature of the PH

19°. A further increase of the prism angle leads to the decrease of the curvature of the PH, and at a prism angle of about 34°, the curvature of the PNJ is almost zero, where the "start point," the "inflection point," and the "end point" of the PNJ form an almost straight line [32].

It is well known that the generation of a strongly localized PNJ depends on the refractive index contrast (RIC) between the dielectric particle and background media and the dimensions of the dielectric particle in the wavelength, rather than the particle 3D shape [10, 27, 36]. However, the formation of a photonic jet through a cuboid particle is due to the phase delay along the wavefront (which represents a concave and convergent wavefront in the cuboid), rather than the factors that govern the focusing by spherical or cylindrical particles [27]. Therefore, the diversity of the thickness with a Janus particle produces an unequal phase of the transmitted plane wave along a particle, resulting in the irregularly concave deformation of the wavefront inside the particle. Interference caused by the diversity of the Janus particle thickness along the polarization direction produces some additional singular points

[37] in the vicinity of this particle, which induces the curvatures appearing at the streamlines. This leads to the formation of a curved localized field near the shadow surface of the Janus particle. It is important to note that in the PH, there is an inflexion point where the curved beam changes its propagating direction. This property is not possessed by the Airy-like beam [38]. Detailed power flow analysis of the wavelength-scaled Janus particles shows [32] that the PH phenomenon is due to the interference of local fields inside the particle and the unequal phase across the particle caused by the particle thickness asymmetry.

1.4 Experimental Verification of the Photonic Hook Phenomenon

Taking into account that the specific value of the incident wavelength is not critical (as long as the mesoscale conditions of particle are satisfied) [36], we showcase a PH effect in terahertz (THz) band (the electromagnetic wave frequency of $\nu = 0.25$ THz or the wavelength of $\lambda = 1.2$ mm, Fig. 1.6) [39–40] using a continuous-wave scanning-probe microscopy based on a flexible 300-μm diameter ($\lambda/4$) bulk sapphire fiber with air cladding as a probe [39–45]. Such waveguides achieve the subwavelength modal confinement in core and provide significant improvement in subwavelength resolution for imaging applications [46–49]. We used polymethyl-pentene (TPX, the refractive index is n = 1.46 at 0.25 THz [50]) as a dielectric material for fabrication of Janus particle. The dielectric structure was fabricated in Tydex

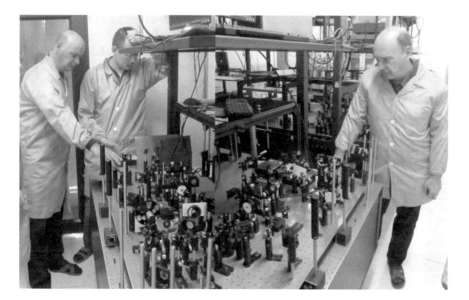

Fig. 1.6 Profs. Igor V. Minin (left) and Oleg V. Minin (right) at the THz laboratory

company [51]. For simulations we used the finite integral technique realized in commercial software – the CST Microwave Studio.

In Fig. 1.7, we illustrate PH evaluation with different dimensions and fixed prism angle. One can see the phenomenon of curved focus bending the Janus particle which is caused by the interference of waves inside it as the phase velocity disperses. Due to the shape of the particle with broken symmetry, the time of the complete phase of the wave oscillations varies irregularly in the particle [32, 39–40], and the emitted electromagnetic beam bends.

An experimental setup based on the principles of fiber-based scanning-probe THz imaging [49] and developed in Bauman Moscow State Technical University is schematically shown in Fig. 1.8.

A backward-wave oscillator [52] equipped with a wire-grid linear polarizer was used as a source of continuous-wave THz radiation. To visualize E^2 field intensity distributions behind the Janus particle, a 2D scanning system with an optical probe made of a flexible fiber with flat ends was used [39–40]. The lateral spatial resolution of the THz imaging was limited by the fiber diameter and was about $\lambda/4$, and the depth resolution was limited by the raster-scan as low as $\lambda/10$. The collimated THz beam radiates the TPX Janus particle from its oblique side. For handling the cuboid inside the THz beam, we glued its flat side onto the 6-μm-thick Mylar film (see insert in Fig. 1.7). The linear polarization of the electric field is directed transversely to the larger Janus particle facet.

The main results of PH visualization are presented in Fig. 1.9. It is worth to note that the PH is formed in the spatial region where the effects of near-field scattering (evanescent fields) play a significant role. Typically, this near-field region is located at the distances not exceeding several particle diameters and is characterized by marked contribution of the radial component of electromagnetic field [27, 36].

From Fig. 1.9 it is followed that the photonic hook's curvature radius is smaller than its operating wavelength and can be adjusted by varying wavelength and geometric parameters of the emitting particle. This is the smallest radius of curvature that's ever been experimentally recorded for electromagnetic waves [39, 40, 53].

It has been shown that the PHs appear in free space behind the dielectric particle but not inside the device "guiding" the light and require no external potentials or waveguiding structures. Although in classical optics it is considered that radiation propagates as straight beams of light, it is possible to produce "curved" rays, exploiting the phase shifts that can occur when light passes through dielectric particles of special shape. The curvature radius and the shape of PH can be adjusted by varying incident light polarization, wavelengths, and geometric and optical parameters of the Janus particle [27, 32, 38–40, 54–59]. PH also has a cross-dimension smaller than the wavelength that makes high resolution possible.

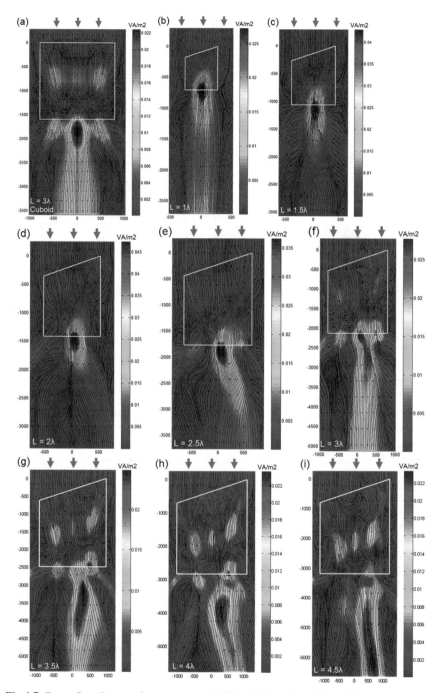

Fig. 1.7 Power flow diagrams for symmetric cuboid $L = 3\lambda$ (**a**) and particles with broken symmetry: $L = 1\lambda$ (**b**), 1.5λ (**c**), 2λ (**d**), 2.5λ (**e**), 3λ (**f**), 3.5λ (**g**), 4λ (**h**), and 4.5λ (**i**) with $\alpha = 18.43°$ (not in scale)

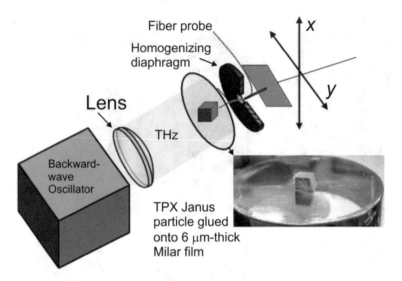

Fig. 1.8 Experimental setup for THz visualization of the photonic hook

1.5   Tailoring "Photonic Hook" from Janus Dielectric Particles with Broken Internal Symmetry

PH formation and its properties by a Janus particle in the form of a cuboid [14–25] with broken shape symmetry [27, 32, 39, 40], which may be considered as the solid immersion lenses and natural Mie scatterers [8, 11], were briefly described above. As it was shown, the localized electromagnetic beam near the rear surface of Janus particle bends due to the constructive interference of waves passing through the asymmetric particle parts with different phase velocities. Recently, for the generation of PHs, the dielectric particle with symmetrical external shape but symmetry-broken material composed with different refractive indexes was proposed [54, 59–60]. Similar particle was numerically studied independently in [61] for solid inorganic and flexible polymer half-cylinders. It could be noted that in the practical applications with dielectric mesoscale particles, their placement on a substrate is usually required. At the same time, it is known that the placement of a cubic particle with a broken external shape symmetry on a dielectric substrate may lead to the PH spoiling because such substrate generates its own scattered field component, which interferes with the PH [15, 62].

Below, following to [56], we discussed the results of investigations of the PH parameters produced by Janus dielectric particles with inhomogeneous internal structure and placed onto an optically transparent substrate, illuminated by a plane wave. New criteria of PH quality for complex beam curvature characterization are also discussed.

The main idea of the PH generation by means of a RIC Janus particle is as follows [56]. Wave phase $\Delta\phi$ control can be carried out both with the help of a particle

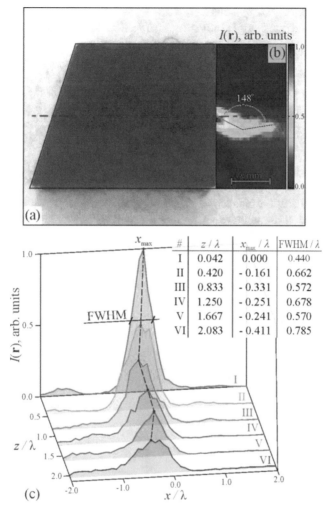

Fig. 1.9 Experimental visualization of PH. (**a**) Janus particle, (**b**) THz field intensity distribution E^2 at the shadow side of a Janus particle, and (**c**) cross-sections of the observed PH. Reprinted from [39] with the permission of AIP Publishing

with asymmetry of the shape (due to different lengths of the particle in its different parts, $\Delta\phi = kn\Delta L$), and with the help of a symmetric particle with built-in asymmetry of the refractive index ($\Delta\phi = k(n_2 - n_1)L$), where k is a wavevector, n is refractive index of the material, and L is path length of the particle. Thus, an optically homogeneous particle with broken symmetry composed by a triangular prism adjacent to the front side of a cuboid particle can be treated as an optically inhomogeneous symmetric Janus particle with rectangular faces.

For characterizing the curvature of the PH, we use the following parameters: the tilt angles of the left α_1 and right α_2 PH arms relative to the direction of wave incidence and the total PH bend angle $\alpha_h = \pi - (\alpha_1 + \alpha_2)$, as well as the subtense L_h of curved beam and hook height increment h, are shown in Fig. 1.10. The slope angle determinations was carried out by a stepwise calculation of the transverse coordinate y_m of field intensity maximum in the field localized region along each hook arm (see Fig. 1.10b).

Let's briefly describe the physical causes of the PH's transverse component of the Poynting vector ($\mathbf{S} = (c/8\pi)\mathbf{E} \times \mathbf{H}$, $\mathbf{E}$ and $\mathbf{H}$ are the electric and magnetic vectors, respectively, c is the speed of light) flow bending during wave diffraction at an anisotropic dielectric composite Janus 2D particle with different RIC $n_2 : n_3$ of the parts (Fig. 1.11).

One can see that the PH is formed mainly by the two most power flows (marked S_1 and S_2 in Fig. 1.11) that emerged in the upper and lower parts of the Janus particle. Due to diffraction on rectangular particle sides, these two power flows are always directed at an angle to each other. Because of their interference forms, a "leaky" external electromagnetic field in the form of a tilted localized field [27, 32, 39, 40] and the resulting localized field area is directed along the direction of illuminating wave ($\alpha_1 = \alpha_2 = 0$). The RIC between the halves of a Janus particle leads to Poynting vector flow redistribution in favor of the flow passing through the lower RIC part (S_1 in Fig. 1.11) due to the action of the fundamental laws of wave refrac-

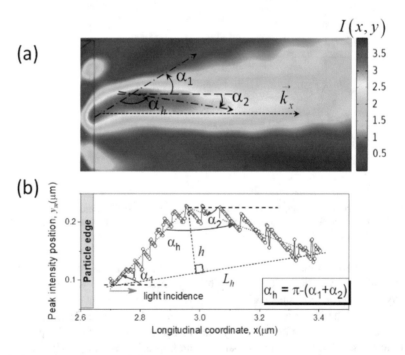

Fig. 1.10 (**a, b**) PH parameters definition: (**a**) normalized localized field intensity profile near the shadow surface of a particle; (**b**) lateral position y_m of the PH intensity maximum versus longitudinal coordinate x. Here, h is the PH height increment, L_h is subtense length. Reprint from [56] with the permission of IOP Publishing

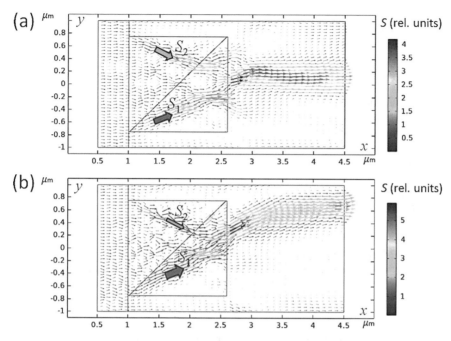

Fig. 1.11 (**a**, **b**) Poynting vector S distribution near the Janus particle on substrate ($n_1 = 1.5$) in air ($n_0 = 1$) with different RICs (**a**) $n_2 : n_3 = 1.65{:}1.5$ ($S_1 \approx S_2$) and (**b**) $n_2 : n_3 = 1.8{:}1.5$ ($S_1 > > S_2$). Reprint from [56] with the permission of IOP Publishing

tion at an interface [63]. As a result, the external field localized area first acquires a "refractive" slope to the rear face of the particle ($\alpha_1 \neq 0$) and then bends ($\alpha_2 \neq 0$) due to interference of the fields in two counterflows. Depending on the ratio of Poynting vector flow intensity, the PH can largely change its bending angle α_h [56].

Simulations show [56] that the increase of substrate RIC with respect to the Janus particle leads to increase of the PH bending angles.

As it was noted above, the unique key characteristics of the PH is the curvature radius, which has subwavelength scales [27, 32, 38–40, 53]. The curvature of the PH can be defined by the height increment h (Fig. 1.10b) as the distance between the center of the PH arc and the center of the subtense. The transversal acceleration of the PH when moving along a curved path in the analogy to the self-accelerated Airy beams [1, 2, 38] corresponds to the height increment and its length. Figure 1.12 shows the dependences of length L_h and height increment h on the Janus particle RIC value n_2. In the case of consideration, the maximum value of h is observed near $n_2 = 1.52$, where the length of the PH is of the order of one wavelength. Importantly, at $n_2 = 1.72$, the height increment h changes sign and becomes negative, which corresponds to PH direction inversion. These properties allow one to implement the dynamic control on the spatial shape of the PH from convex to concave by changing the RIC n_2 of the Janus particle material.

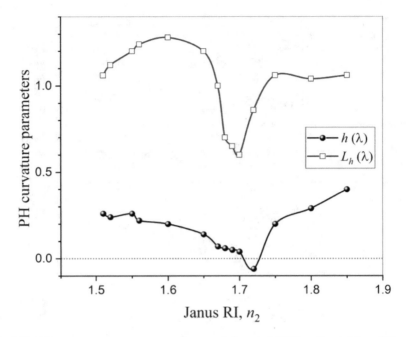

Fig. 1.12 PH curvature parameters versus Janus particle material RIC n_2. Reprint from [56] with the permission of IOP Publishing

1.6   Formation of the "Photonic Hook" from Janus Cylindrical Particles with Broken Internal Symmetry

As it was mentioned above, for the generation of PHs, the dielectric particle with symmetry-broken material composed with different RICs was proposed in [54, 59–60] and independently in [61] for solid inorganic and flexible polymer half-cylinders studied. By varying the RIC of the two half-cylinders, the bending degrees and inflection point of the PHs can be effectively controlled. To this end, the localized field area formation near the shadow surface of the half-cylinder [64], which were previously studied, is of great interesting to understand the nature of the PH formation.

Figure 1.13 demonstrates the dynamics of the localized fields formation near the shadow surface of the cylindrical Janus particle with refractive indexes of the half-cylinders as n = 1.3 (upper half-cylinder) and n = 1.5 (bottom half-cylinder) vs rotation angle for TE and TM modes. The occurrence of reflection (including total internal reflection) and refraction and the backscattering effect are clearly visible. Note that in units normalized to the incident field, the formed curvilinear localized beams are more "magnetic" than "electric" [10, 65–67].

The similar effect based on Janus hollow microcylinder liquids-filled by two insoluble liquids, and which illuminated by a plane wave to produce photonic hook, was considered in [68]. Notice that the idea of using two insoluble liquids to control a localized field was discussed earlier in acoustics [69]. It has been shown that the

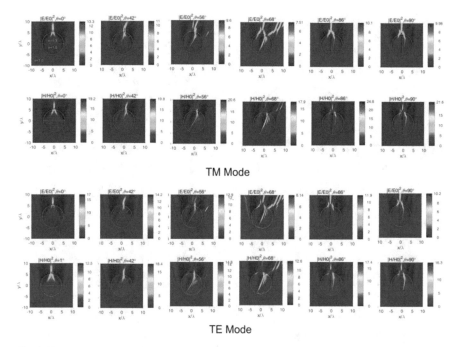

TM Mode

TE Mode

Fig. 1.13 Localized field formation near the shadow surface of the cylindrical Janus particle with refractive indexes of the half-cylinders as n = 1.3 (upper half-cylinder) and n = 1.5 (bottom half-cylinder) vs rotation angle for TE and TM modes

profile and properties of the photonic hook can be tuned by rotating the hollow microcylinder or changing the light incident angle due to the Poynting vector, and energy flow distribution is inhomogeneous on the two sides of the transmission axis (Fig. 1.14).

1.7   Formation of the Optical Clamp (Twin "Photonic Hook")

The interaction of a single plane wave with structured particle allows formation of a twin photonic hook. In [70] the formation of twin photonic hooks with a single illuminating light beam and a symmetric mesoscale particle in the form of combined twin-ellipse cylinders were analyzed. Such twin-ellipse cylindrical particle was created by overlapping two cylinders at their semi-minor axis, while the semi-major axis remains the same. Under the illumination of plane wave, this combined particle can generate twin photonic jets and bend them to form twin photonic hooks.

The use of two plane wave beams to illuminate a mesoscale sphere and cylinder to form twin photonic jets in application to nanoparticle manipulation was considered in [71]. Later, the generation of twin photonic hook (Fig. 1.15a) by a single

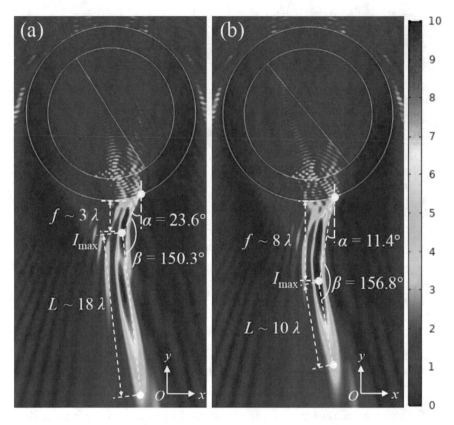

Fig. 1.14 The field intensity distribution, decay length L, focal distance f, tilt angle-α. and bending angle-β for the cases of rotation angles (**a**) 33° and (**b**) 40°. Reprinted from [68] with permission by Prof. Guoqiang Gu, Department of Electrical and Electronic Engineering, Southern University of Science and Technology, China

cylindrical particle under structured illumination made up of two crossing opposite polarized plane waves was considered in [72]. It has been shown that the Poynting vector flow in a cylindrical particle produced PNJ under single TE-polarization plane wave is symmetric relative to axis of symmetry. At the same time, the Poynting vector flow under two opposite polarized plane waves with the relative big offset angle is bending to form the twin PHs. Simulations of the 5 μm in diameter cylindrical particle (under two opposite polarized plane waves with the offset angle of 60°) have shown that the FWHM of the PHs is bigger than half of the incident wavelength with curvature of the PH about 24°.

Two adjacent parallel cylinders illuminated by a single TE-polarization plane wave may apply to observe the similar effect [73]. It has been shown that the shape profiles of the PHs can be controlled by changing the gap between the two cylinders [74] and diameters (Fig. 1.15b). The curvature of the PH decreases with decreasing

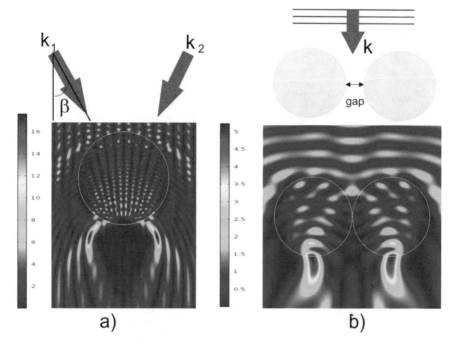

Fig. 1.15 Photonics clamp in the form of twin optical hooks [71, 72]: (**a**) under two opposite polarized plane waves with $\beta = 30°$, refractive index contrast 1.46, Mie size parameter $q = 28.55$ and (**b**) under the plane wave with gap = 0.028λ, Mie size parameter $q = 7.42$, and refractive index contrast 1.46. Not in scale. With permission by Song Zhou, Faculty of Mechanical and Material Engineering, Huaiyin Institute of Technology, China

the gap between cylinders. At the same time, the FWHM of the PH is larger than that of the PNJ formed by a single cylinder but less than the diffraction limit.

Conclusion PH can be created using a compact mesoscale Janus near its shadow surface in contrast to classical Airy-family beams, generated using a spatial light modulator behind the focus of spherical lens or by a complicated optical element with a cubic phase. Moreover, in the case of PH, only the main lobe has a curved shape, while the family of curved sidelobes is absent in contrast to Airy-like beam, which consists of a main lobe and a family of sidelobes, whose intensity decays exponentially [53], and the FWHM of PH is smaller than that for PNJ and far less than the diffraction limit. It is essential that the simplicity of the PH source and its wavelength-scale dimensions allow it to be integrated into lab-on-chip platforms.

The shape of the photon hook and the characteristics of curved near-field structured beams can be quite exotic due to the control of phase delays across the wavefront by choosing the shape of mesoscale particles. Figure 1.16 has shown the exotic looped trajectories of photonic flow.

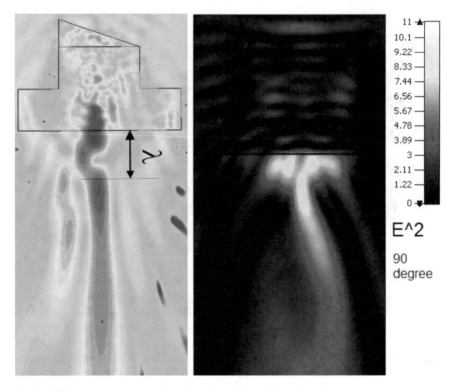

Fig. 1.16 Example of a subwavelength photonic loop [38] (left) and photonic hook at the side illumination [54] (right, available via license: CC BY 4.0).

The unique fantastic qualities of subdiffraction limited structured beam waists and subwavelength-scaled curvature radius show important application prospects in the fields of imaging, nano-manipulation, nonlinear optics, integrated systems, etc. [11, 38, 53, 59].

References

1. G. Siviloglou, J. Broky, A. Dogariu, D.N. Christodoulides, Observation of accelerating Airy beams // Phys. Rev. Lett., 99, 213901 (2007).
2. N. Efremidis, Z. Chen, M. Segev, D.N. Christodoulides, Airy beams and accelerating waves: An overview of recent advances // Optica 6, 686–701 (2019).
3. L. Froehl, L. Froehly, F. Courvoisier, A. Mathis, M. Jacquot, L. Furfaro, R. Giust, P. Lacourt, J. Dudley, Arbitrary accelerating micron-scale caustic beams in two and three dimensions // Opt. Express 19, 16455–16465 (2011).
4. N. Gao, H. Li, X. Zhu, Y. Hua, G. Xie, Quasi-periodic gratings: Diffraction orders accelerate along curves // Opt. Lett. 38, 2829–2831 (2013).

5. J. He, S. Wang, Z. Xie, J. Ye, X. Wang, Q. Kan, Y. Zhang, Abruptly autofocusing terahertz waves with meta-hologram // Opt. Lett. 41, 2787–2790 (2016).
6. W. Chan, H. Chen, A. Taylor, I. Brener, M. Cich, D. Mittleman, A spatial light modulator for terahertz beams // Appl. Phys. Lett. 94, 213511 (2009).
7. A. Valdmann, P. Piksarv, H. Valtna-Lukner, P. Saari, White-light hyperbolic Airy beams // J. Opt. 20, 095605 (2018).
8. A. Heifetz, S. Kong, A. Sahakian, A. Taflove, V. Backman, Photonic nanojets // J. Comput. Theor. Nanosci. 6, 1979–1992 (2009).
9. C. Liu, Y. Wang, Real-space observation of photonic nanojet in dielectric microspheres // Physica E 61, 141–147 (2014).
10. B. Luk'yanchuk, R. Paniagua-Domínguez, I.V. Minin, O.V. Minin, Z. Wang, Refractive index less than two: Photonic nanojets yesterday, today and tomorrow (Invited) // Opt. Mater. Express 7, 1820–1847 (2017).
11. I.V. Minin, C.-Y. Liu, Y. E Geints, O. V. Minin. Recent advantages in Integrated Photonic Jet-Based Photonics // Photonics 7(2), 41 (2020)
12. Y. Kopylov, A. Popov, Diffraction phenomena inside thick Fresnel zone plates // Radio Sci. 31, 1815–1822 (1996).
13. Born, M.; Wolf, E. *Principles of Optics*, 6th ed.; Pergamon: Oxford, UK, 1986.
14. V. Pacheco-Peña, M. Beruete, I.V. Minin, O.V. Minin, Terajets produced by dielectric cuboids. // Appl. Phys. Lett. 105, 084102 (2014).
15. O.V. Minin, I.V. Minin, Terahertz artificial dielectric cuboid lens on substrate for super-resolution images. // Opt. Quantum Electron. 49, 326–329 (2017).
16. I.V. Minin, O. V. Minin, V. Pacheco-Peña, M. Beruete. All-dielectric periodic terajet wave-guide using an array of coupled cuboids // Applied Physics Letters, 106, 254102 (2015)
17. M. Khodzinsky, A. Vosianova, V. Gill, A. Chernyadiev, A. Grebenchukov, I. V. Minin, and O. V. Minin. "Formation of terahertz beams produced by artificial dielectric periodical structures", Proc. SPIE 9918, Metamaterials, Metadevices, and Metasystems 2016, 99182X (16 September 2016)
18. I.V. Minin, O.V. Minin, I. Nevedof, V. Pacheco-Peña, M. Beruete. Beam compressed system concept based on dielectric cluster of self-similar three-dimensional dielectric cuboids // An International Joint Conference of The 9th Global Symposium on Millimeter-Waves (GSMM 2016) and The 7th ESA Workshop on Millimetre-Wave Technology and Applications, June 6–8, 2016, Aalto University, Espoo 64–66
19. H. H. Pham, S. Hisatake, T. Nagatsuma, I. V. Minin and O. V. Minin, "Experimental characterization of terajet generated from dielectric cuboid under different illumination conditions," // *2017 10th Global Symposium on Millimeter-Waves*, Hong Kong, 2017, pp. 120–122
20. L. Yue, B. Yan, J. Monks, Z. Wang, N. T. Tung, V. D. Lam, O. V. Minin, and I. V. Minin. Production of photonic nanojets by using pupil-masked 3D dielectric cuboid // J. Phys. D: Appl. Phys. 50, 175102 (2017)
21. C.-Y. Liu, T.-P. Yen, O. V. Minin and I. V. Minin. Engineering photonic nanojet by a graded-index micro-cuboid. // Physica E: 98 105–110 (2018)
22. L. Yue, B. Yan, J. Monks, Z. Wang, N. T. Tung, V. D. Lam, O. V. Minin, and I. V. Minin. A millimetre-wave cuboid solid immersion lens with intensity-enhanced amplitude mask apodization. // Journal of Infrared, Millimeter, and Terahertz Waves, 39(6), 546–552 (2018)
23. Y. E. Geints, O. V. Minin and I. V. Minin. Systematic study and comparison of photonic nanojets produced by dielectric microparticles in 2D- and 3D- spatial configurations. // Journal of Optics 20, 065606 (2018)
24. H. Pham, S. Hisatake, I. V. Minin, O. V. Minin, and T. Nagatsuma. Three-Dimensional Direct Observation of Gouy Phase Shift in a Terajet Produced by a Dielectric Cuboid // APL, 108, 191102 (2016)

25. H. Pham, S. Hisatake, O.V. Minin, T. Nagatsuma and I.V. Minin. "Asymmetric Phase Anomaly of Terajet Generated from Dielectric Cube under Oblique Illumination." // Appl. Phys. Lett. 110(20), 201105 (2017)

26. Gouy, L. G. Sur une propriete nouvelle des ondes lumineuses// *C. R. Acad. Sci. Paris* 110, 1251 (1890).

27. I. V. Minin, O. V. Minin, *Diffractive Optics and Nanophotonics: Resolution Below the Diffraction Limit*, Springer, Cham (2015).

28. Y. Cao, Z. Liu, O. V. Minin and I. V. Minin. Deep subwavelength-scale light focusing and confinement in nanohole-structured mesoscale dielectric spheres // *Nanomaterials*, *9(2)*, 186 (2019)

29. I. V. Minin, O. V. Minin, Y. Cao, Z. Liu, Y. Geints and A. Karabchevsky. Optical vacuum cleaner by optomechanical manipulation of nanoparticles using nanostructured mesoscale dielectric cuboid // Sci Rep *9*, 12748 (2019)

30. O. V. Minin and I. V. Minin. Ultrafast all-optical THz modulation based on wavelength scaled dielectric particle with graphene monolayer // Proc. SPIE 11065, Saratov Fall Meeting 2018: Optical and Nano-Technologies for Biology and Medicine, 110651J (3 June 2019); https://doi.org/10.1117/12.2525077

31. Houston WV. A compound interferometer for fine structure work // Phys Rev 29, 0478–0484 (1927)

32. L. Yue, O. V. Minin, Z. Wang, J. Monks, A. Salin, and I. V. Minin, Photonic hook: a new curved light beam // Optics Letters 43(4), 771–774 (2018).

33. I. H. Malitson, Refractive-Index Interpolation for Fused Silica // J. Opt. Soc. Am. 55, 1205 (1965).

34. E. Abbe, Beiträge zur Theorie des Mikroskops und der mikroskopischen Wahrnehmung // Archiv für Mikroskopische Anatomie, **9**, pages 413–468 (1873); https://doi.org/10.1007/BF02956173

35. Lord Rayleigh F.R.S. XXXI. Investigations in optics, with special reference to the spectroscope // The London, Edinburgh, and Dublin Philosophical Magazine and Journal of Science, 8(49), 261–274 (1879); https://doi.org/10.1080/14786447908639684

36. I.V. Minin, O.V. Minin, Y. Geints, Localized EM and photonic jets from non-spherical and non-symmetrical dielectric mesoscale objects: Brief review // Annalen der Physik 527(7–8), 491–497 (2015)

37. Z. B. Wang, B. S. Luk'yanchuk, M. H. Hong, Y. Lin, and T. C. Chong, Energy flow around a small particle investigated by classical Mie theory// Phys. Rev. B 70, 035418 (2004).

38. I. V. Minin and O. V. Minin. Recent Trends in Optical Manipulation Inspired by Mesoscale Photonics and Diffraction Optics // J of Biomedical Photonics & Eng 6(2), 020301 (2020)

39. I. V.Minin, O. V. Minin, G. Katyba, N. Chernomyrdin, V. Kurlov, K. I. Zaytsev, L. Yue, Z. Wang, and D. N. Christodoulides. Experimental observation of a photonic hook // Appl. Phys. Lett. 114, 031105 (2019)

40. O. V. Minin, I. V. Minin, K. I. Zaytsev, G. Katyba, V. Kurlov, L. Yue, Z. Wang, "Electromagnetic field localization behind a mesoscale dielectric particle with a broken symmetry: a photonic hook phenomenon," Proc. SPIE 11368, Photonics and Plasmonics at the Mesoscale, 1136807 (2 April 2020)

41. S. Jamison, R. McGowan, and D. Grischkowsky, Single-mode waveguide propagation and reshaping of sub-ps terahertz pulses in sapphire fibers // Applied Physics Letters, 76(15), 1987–1989 (2000).

42. A. Barh, B. P. Pal, G. P. Agrawal, R. K. Varshney and B. M. A. Rahman, "Specialty Fibers for Terahertz Generation and Transmission: A Review," // *IEEE Journal of Selected Topics in Quantum Electronics*, 22(2), 365-379 (2016).

43. G. M. Katyba, N. V. Chernomyrdin, I. N. Dolganova, A. A. Pronin, I. V. Minin, O. V. Minin, K. I. Zaytsev, and V. N. Kurlov. "Step-index sapphire fiber and its application in a terahertz near-field microscopy", Proc. SPIE 11164, Millimetre Wave and Terahertz Sensors and Technology XII, 111640G (18 October 2019)

44. N. Chernomyrdin, G. Katyba, A. Gavdush, T. Frolov, I. Dolganova, V. Kurlov, and K. I. Zaytsev. "Terahertz transmission-mode near-field scanning-probe microscope based on a flexible sapphire fiber", Proc. SPIE 11088, Optical Sensing, Imaging, and Photon Counting: From X-Rays to THz 2019, 110880I (9 September 2019)

45. Md. S. Islam, C. M. B. Cordeiro, M. A. R. Franco, J. Sultana, A. L. S. Cruz, and D. Abbott. Terahertz optical fibers // **Optics Express** 28 (11), 16089–16117 (2020)

46. N. Chernomyrdin, A. Kucheryavenko, G. Kolontaeva, G. Katyba, P. Karalkin, V. Parfenov, A. Gryadunova, N. Norkin, O. Smolyanskaya, O. V. Minin, I. V. Minin, V. Karasik, K. I. Zaytsev, "A potential of terahertz solid immersion microscopy for visualizing subwavelength – scale tissue spheroids," Proc. SPIE 10677, Unconventional Optical Imaging, 106771Y (24 May 2018)

47. W. Talataisong, J. Gorecki, R. Ismaeel, M. Beresna, D. Schwendemann, V. Apostolopoulos & G. Brambilla, Singlemoded THz guidance in bendable TOPAS suspended-core fiber directly drawn from a 3D printer // Sci Rep **10,** 11045 (2020)

48. C.-M. Chiu, H.-W. Chen, Y.-R. Huang, Y.-J. Hwang, W.-J. Lee, H.-Y. Huang, and C.-K. Sun, All-terahertz fiber-scanning near-field microscopy // Opt. Lett. 34, 1084 (2009).

49. I. V. Minin and O. V. Minin. "System of microwave radiovision of three-dimensional objects in real time", Proc. SPIE 4129, Subsurface Sensing Technologies and Applications II, (6 July 2000)

50. A. Podzorov and G. Gallot, Low-loss polymers for terahertz applications // Appl. Opt. 47, 3254 (2008).

51. Tydex: http://www.tydexoptics.com/products/thz_optics/thz_materials/

52. A. Dobroiu, M. Yamashita, Y. N. Ohshima, Y. Morita, C. Otani, and K. Kawase. Terahertz imaging system based on a backward-wave oscillator // Appl. Opt., 43(30), 5637–5646 (2004)

53. K. Dholakia, G. Bruce, Optical hooks // Nature Photonics 13(4), 229–230 (2019).

54. I. V. Minin, O. V. Minin, L. Yue, Z. Wang, V. Volcov, and D. N. Christodoulides. Photonic hook – a new type of subwavelength self-bending structured light beams: a tutorial review // ArXiv: 1910.09543 (2019)

55. I. V. Minin, C.-Y. Liu, Y.-C. Yang, K. Staliunas and O. V. Minin. Experimental observation of flat focusing mirror based on photonic jet effect // Sci Rep 10, 8459 (2020).

56. Y. Geints, I. V. Minin, O. V. Minin. Tailoring "photonic hook" from Janus dielectric microbar // Journal of Optics, 22, 065606 (2020)

57. C.-Y. Liu; H.-J. Chung, O.V. Minin, I.V. Minin. Shaping photonic hook via well-controlled illumination of finite-size graded-index micro-ellipsoid // Journal of optics 22(8), 085002 (2020)

58. I. V. Minin, O. V. Minin, C.-Y. Liu, and H.-D. Wei, Y.Geints, A. Karabchevsky. Experimental demonstration of tunable photonic hook by partially illuminated dielectric microcylinder // Optics Letters 45(17), 4899–4902 (2020)

59. I.V. Minin, O. V. Minin. Dielectric particle-based strategy to design a new self-bending subwavelength structured light beams. // IOP Conf. Ser.: Mater. Sci. Eng. 1019, 012093 (2021)

60. I.V.Minin and O.V.Minin. "Subwavelength self-bending structured light beams." In: Proc. of the Fourth Russian-Belarusian Workshop "Carbon nanostructures and their electromagnetic properties", Tomsk, Apr. 21–24, 2019. P.52–57.

61. G. Gu, L. Shao, J. Song, J. Qu, K. Zheng, X. Shen, Z. Peng, J. Hu, X.Chen, M. Chen, and Q. Wu, "Photonic hooks from Janus microcylinders," // Opt. Express 27(26), 37771–37780 (2019)

62. A.S. Ang, I.V. Minin, O. V. Minin, S.V. Sukhov, A. Shalin, A. Karabchevsky. "Low-contrast photonic hook manipulator for cellular differentiation." Proc. of the 9th Int. conf. on Metamaterials, Photonic crystals and Plasmonics (2018).

63. N. Yu, P. Genevet, M. Kats, F. Aieta, J. Tetienne, F. Capasso, and Z. Gaburro. Light Propagation with Phase Discontinuities: Generalized Laws of Reflection and Refraction // Science 334, 333 (2011).

64. I. Mahariq, I. Giden, H. Kurt, O. V. Minin and I. V. Minin. Strong electromagnetic field local-ization near the surface of hemicylindrical particles // Opt Quant Electron 49, 423 (2017)
65. Z. Wang, B. Luk'yanchuk, L. Yue, B. Yan, J. Monks, O. V. Minin, I. V. Minin, S. Huang and A. A. Fedyanin. High order Fano resonances and giant magnetic fields in dielectric micro-spheres // Sci Rep 9, 20293 (2019)
66. L. Yue, B. Yan, J. N. Monks, R. Dhama, C. Jiang, O. V. Minin, I. V. Minin, and Z. Wang. Full three-dimensional Poynting vector analysis of great field-intensity enhancement in a specifi-cally sized spherical-particle // Sci Rep 9, 20224 (2019)
67. L. Yue, Z. Wang, B. Yan, J. Monks, Y. Joya, R. Dhama, O. V. Minin, and I. V. Minin. Super-enhancement focusing of Teflon sphere // Annalen der Physik 2000373 (2020)
68. Z. Peng, G. Gu, L. Shao, X. Shen. Easily tunable long photonic hook generated from Janus liquids-filled hollow microcylinder // arXiv:2007.13093 (2020)
69. S. Perez-Lopez, P. Candelas, J. M. Fuster, C. Rubio, O. V. Minin, and I. V. Minin, Liquid-liquid core-shell configurable mesoscale spherical acoustic lens with focusing below the wave-length // Appl. Phys. Express 12, 087001 (2019).
70. X. Shen, G. Gu, L. Shao, Z. Peng, J. Hu, S. Bandyopadhyay, Y. Liu, J. Jiang, and M. Chen. Twin photonic hook generated by twin-ellipse microcylinder // IEEE Photonics Journal, 12(3), 1–9, Art no. 6500609 (2020)
71. A. Poteet, X. Zhang, H. Nagai and C. Chang. Twin photonic nanojets generated from coherent illumination of microscale sphere and cylinder, // Nanotechnology, 29 (7), 075204 (2018).
72. S. Zhou. Twin photonic hooks generated from two coherent illuminations of a micro-cylinder// J. Opt. 22, 085602 (2020)
73. S. Zhou. Twin photonic hooks generated from two adjacent dielectric cylinders. // Opt Quant Electron 52, 389 (2020).
74. E. Bulgakov, K. Pichugin, A. Sadreev, Evolution of the resonances of two parallel dielectric cylinders with distance between them // Phys. Rev. A 100, 043806 (2019)

Chapter 2
Formation of a Photon Hook by a Symmetric Particle in a Structured Light Beam

Abstract In this chapter, we briefly discuss the observations of a tunable photonic hook generated by symmetric microparticles in a structured light beam. Partial irradiation of a dielectric symmetric particle with a wide beam of light may be considered as a simple example of the structured light beams. A moveable opaque metal mask was mounted in front of the cylindrical and graded oblate spheroid implementing such a partial illumination and imparting spatial curvature to the photonic hook. It has been shown that the key parameters of the photonic hook may be controlled by both the width of illuminated beam and it positions and polarization state.

2.1 The Photonic Hook Formation Upon Asymmetric Illumination of a Symmetric Mesoscale Particle

The influence of specific illumination conditions (coaxial illumination with adjustable area and boundary illumination) for spherical BaTiO3 particles with diameter approximately 127λ ($q \sim 42\pi$) was considered for an optical image contrast [1]. A multifocal curved beam based on off-axis incidence of Gaussian beam for SiO_2 microsphere with a diameter of 433λ ($q \sim 144\pi$, i.e., in geometrical optics approximation) was considered in the optical waveband [2]. However, direct experimental observation of the photonic hook was not performed under the conditions of mesoscale particles for the photonic nanojet (PNJ) effect. The PH through a glass cuboid embedded in a structured dielectric cylinder is discussed by numerical calculations [3]. Furthermore, two PHs are observed by specially designed five-layer dielectric cylinder [4]. But most of these known solutions require specially designed particles, which limits the further development of this type of curved beams. The anomalously intensity-enhanced apodization effect based on axial centered amplitude mask was discovered of focusing on the different types of particle lenses in [5–7] in application to the PNJ beamwidth.

Below we briefly describe the experimental direct imaging of the PH created by mesoscale cylindrical particle ($q \sim 12\pi$) with partial blocking by an asymmetric amplitude mask [8].

Figure 2.1a illustrates the proof of the concept of the problem, and Fig. 2.1b illustrates the definition of curvature for a PH [9–12]. The production capability of

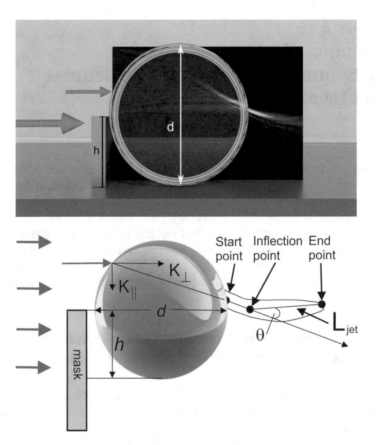

Fig. 2.1 The proof of the concept of the PH generation (**a**) and the definition of curvature for a PH (**b**)

curved PNJ was numerically studied using a dielectric microcylinder illuminated by a laser beam under different heights of the metallic mask. The incident Gaussian beamwidth was 1 mm at the wavelength of $\lambda = 405$ nm with the 5 µm diameter of dielectric microcylinder. The metallic mask was placed near the cylinder surface less than 0.7 µm. The curvature of the PH is defined by the tilt angle θ [9–12] as the angle between the propagation direction and horizontal axis (x axis). We use the finite-difference time-domain (FDTD) method with perfectly matched layer boundary conditions in simulations.

The field intensity distributions at several key h values by numerical simulations are illustrated in the left column of Fig. 2.2 to reveal the curvature changes of the PHs. The case of $h = 0$ corresponds to the classical PNJ for dielectric microcylinder. As shown in Fig. 2.2, the PH's shape and curvature radius can be adjusted by varying mask height. When the axis of the spatially limited light beam is shifted to the cylindrical particle edge, the optical axis of PNJ deflects due to the curvature of the particle cylindrical surface. The length of PNJ decreases with a decrease of its inten-

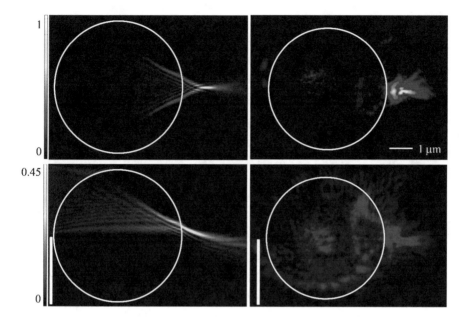

Fig. 2.2 Numerical results (left column) and experimental raw images (right column) of normalized intensity maps for the dielectric microcylinder with metallic mask at (**a**) $h = 0$ and (**b**) $h = 0.5d$. Adapted from [13]

sity, and the orientation of PNJ deflects toward the main axis of the particle. The simulation results show that it is possible to control the curvature of the beam by adjusting the relative position between an off-axis beam and a cylindrical or spherical particle. This phenomenon is similar to cylindrical lens with a decentered aperture under normal illumination because of the large spherical aberration [14, 15].

The physics of the PH formation in this case can be explained as follows [8]. The width of illumination and the refractive index of cylinder determine the angle of refraction on an interface via the generalized Snell's law [16]. Part of the illuminating beam, determined by the mask height, is refracted first on the front surface of the cylinder. Then, the light beam inside the cylinder is refracted a second time when exits from the shadow surface of the cylinder. If the width of the illuminating beam is less than the cylinder diameter, the components of the wave vector $K_{\parallel}$ do not cancel each other through local destructive interference [17]. The wave vector $K_{\parallel}$ is relative to the axis of symmetry of the cylinder which leads to the PNJ curvature. On the other hand, the components of the wave vector $K_{\perp}$ determine the length of the PNJ along the propagation direction. Therefore, the local interference of the optical fields inside the cylinder can generate the PH depending on the width of illuminating beam.

It is interesting to note that the distribution of the field intensity near the shadow surface of the cylinder across the optical axis is close (at lease of the first sidelobe) to the distribution of the Airy function – see Fig. 2.3.

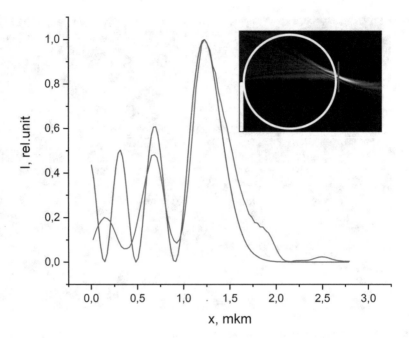

Fig. 2.3 The field intensity distribution near the shadow surface of the cylinder across the optical axis (along a red line shown in the insert), red line, and the Airy function $Ai^2(x)$ (black line). The mask height is $h = 0.5d$

The experimental setup is described in detail in [8]. In the experiment, the polydimethylsiloxane (PDMS) microcylinder was illuminated by a coherent laser beam along the x direction. We assemble the scanning optical microscope to capture the experimental raw images [18]. The scanning in z direction is performed for obtaining clear raw images of the PHs [19]. In the right column of Fig. 2.2, we show the experimental results of PH visualization. In both simulations and experiments, the variation of PH curvature depends on the mask height. By changing the mask height, the localized field (PH) is curved to the y direction.

It could be noted that, in common case, the tilt angle θ calculation methodic is as follows. The primary data is the 2D distribution of PH intensity recorded in the experiments (Fig. 2.4), which is first digitized to outline the hook body at the I_{max}/e level by means of any software. Then this body is divided into two parts representing left and right arms of the hook by drawing the vertical line through the inflection point having maximal (I_{max}) intensity. The start and end points of the hook are selected as the extreme points of both the left and right arms, respectively.

Figure 2.5 summarizes the PH curvature and the peak intensity compared to these in the full illumination case ($h = 0$). In order to evaluate the FWHM of PH, we find the start point and the inflection point (peak amplitude) in the experimental images. The propagation direction is from the start point to the inflection point as shown in Fig. 2.1. The FWHM is the double distance perpendicular to the propaga-

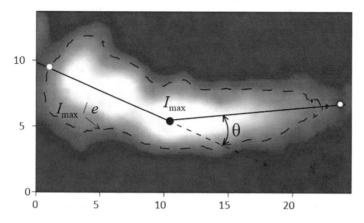

Fig. 2.4 Experimental PH morphology: 2D distribution of PH intensity (zoom image from Fig. 2.2), where PH body is shown by the dashed curve and the tilt angle is denoted as θ

tion direction between the maximal peak amplitude and half-maximum point. In Fig. 2.5, the diffraction limited PH (FWHM $\leq \lambda/2$) is observed at mask height $h \leq 0.15d$ with small tilt angle less 2.5°, while the maximal field intensity decreases by up to 10%. But FWHM of the PH is more than that of PNJ (at $h = 0$) at $h > 0.15d$ as well. In Figs. 2.2 and 2.5, it is clear to see that the mask height has a significant influence to curvature of the PH (tilt angle). When h increases, the curvature degree θ is on a rising trend. It can be seen that the tilt angle θ of the PH is about 20° at the mask height of $h = 0.5d$. The full length of the PH in the case of $h = 0.5d$ is about 3 μm in which the distance between start point and inflection point is 0.8 μm and the distance between inflection point and end point is 2.2 μm. In the case of absence of a mask, the FWHM of the PNJ is less than the half of wavelength. However, the FWHM of the PH is 0.76λ when the mask height is equal to the half of the cylinder diameter. In Fig. 2.5c, the peak intensity of the PH decreases as the mask height increases. At $h = 0.5d$, the peak intensity has decreased to approximately 0.5, half of its original intensity.

In order to demonstrate the curvature of the PH, the experimental cross-sections of the observed curved focus at $h = 0.25d$ and $h = 0.5d$ are shown in Fig. 2.6. Serial transverse cross-sections correspond to serial focal planes along the x axis moving upward by steps of 200 nm. We acquire the cross-sections from the raw images in Fig. 2.2 and select five cross-sections from the inflection point for a better observation, as schematically shown in the insert of Fig. 2.6.

In general, the classical PNJ is formed from a dielectric microcylinder (see Fig. 2.2a). When the incident beam illuminates a full surface of the microcylinder, optical beam will divide into two branches inside the microcylinder and converge into a PNJ near the shadow surface of the microcylinder [3, 4, 19–21]. For a masked microcylinder, one of the branches inside the microcylinder is blocked by the metallic mask as shown in Fig. 2.2b. Only one part of optical beam inside the microcyl-

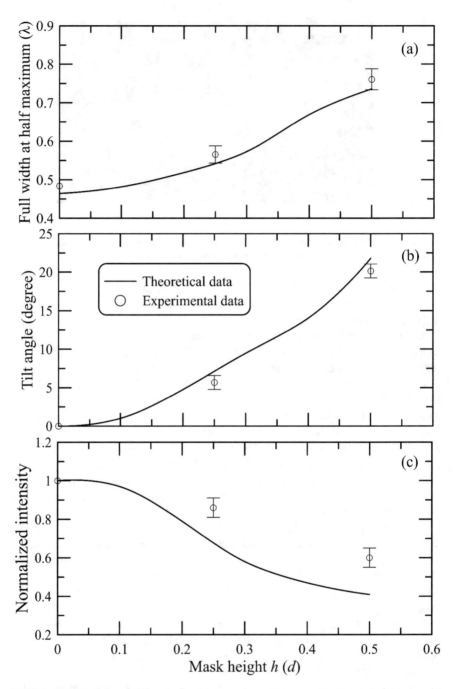

Fig. 2.5 (a) Full width at half maximum, (b) tilt angle, and (c) normalized intensity of the curved focus as a function of the mask height [13]

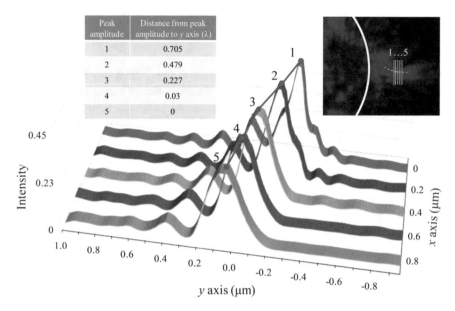

Peak amplitude	Distance from peak amplitude to y axis (λ)
1	0.705
2	0.479
3	0.227
4	0.03
5	0

Fig. 2.6 Experimental cross-sections of the observed curved focus at $h = 0.5d$. The insert indicates the positions of serial cross-sections. The table inset shows the offset distance from each peak amplitude to y axis. Adapted from [13]

inder is refracted on the rear surface of the microcylinder, and then a curved focus area is formed near the shadow surface of the microcylinder.

On the other words, the time of a complete phase cycle of the wave oscillations inside the microcylinder is irregular due to asymmetry of illuminating wave by amplitude mask apodization, and this is the cause of bending of the output PH. The table inset in Fig. 2.6 shows the offset distance from each peak amplitude to y axis. It is clearly seen that the position of the peak amplitude in the region of radiation localization shifts in the transverse direction (y axis) when the distance from the shadow surface of the microcylinder increases in the horizontal direction (x axis). In other words, the localized optical beam is not direct, and the focusing beam is clearly bent to form a PH near the rear surface of the microcylinder. As it follows from the experiments and simulations, the offset distance of the PH increases 5.5 times in the 1 μm length with increasing mask height from $h = 0.25d$ to $h = 0.5d$ (Fig. 2.6). Thus, increasing the mask height leads to a significant increase in the PH curvature. In terms of tilt angle, a larger h brings a bigger θ. However, the price of increasing the PH curvature is a drop in the maximum intensity of the focus, as shown in Figs. 2.5 and 2.6. Compared to a conventional PNJ, the PH should cause an increase in spatial resolution and field of view.

2.2   Shaping Photonic Hook Upon Structured Illumination of a Graded-Index Ellipsoid

The use of the multilayered graded-index particle is an effective way to increase the length of classical PNJ [23–30]. In Ref. [31] the formation of a long photonic hook created by mesoscale graded-index oblate spheroid (elliptical particle) with partially blocking by an amplitude mask was described. Parameters of the dielectric particle, which consist of a core and four concentric layers with equal thickness, were selected as follows: the minor and major radii of the oblate spheroid $a = 1.0$ μm (1.88λ) and $b = 3.0$ μm (5.64λ), respectively, where λ is the incident wavelength of 0.532 μm. The number of shells is m, and the refractive index of every layer is defined by n_m ($m = 0$–4).

Figure 2.7 illustrates the simulation model of the multilayer oblate spheroid with metallic mask. The refractive index contrast from layer to layer of the graded-index oblate micro spheroid can be expressed as $n_m/n_c = (n_s/n_c)^{(m/s)g}$ [24], where g is the index grading type parameter, $m = 0$ is the dielectric core ($n_c = 1.5$), and $m = 4$ is the outer layer ($n_s = 1.1$). The refractive index grading has a minimum value ($n_s = 1.1$)

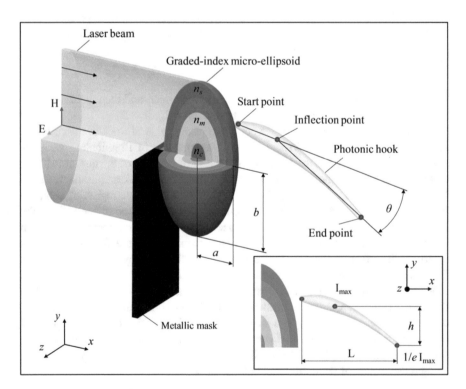

Fig. 2.7 Schematic diagram of the graded-index oblate micro spheroid with metallic mask for photonic hook formation. The insert indicates the curvature definition of photonic hook similar to Fig. 2.1. Reprinted from [31] with permission by IOP Publishing

at the outer layer in the radial direction and a maximum value ($n_c = 1.5$) at the dielectric core. Three types of the index grades are defined by the g value: the convex, concave, and linear. If g is more than 0, the refractive index grading (RIG) starts from the densest core and terminates in the outermost layer with lowest refractive index. If g is equal to 1, the RIG is the linear layer-by-layer variation with the constant contrast. If g is much larger than 1, the multilayer ellipsoid becomes similar to a homogeneous oblate micro spheroid with high refractive index [24].

Simulations show in [31] that the PH curvature increases as the mask height increases, similar to the case of a homogeneous particle. However, the PH focus (start and inflection points) moves inside the oblate micro spheroid when the height of the amplitude mask is 1.25b. The smallest PH curvature observed when the mask height is 0.5b.

Figure 2.8 shows the power flow distributions of the PHs formed in the vicinity of elliptical particle with fixed mask height at different index grading type parameters. The analysis of the results from Fig. 2.8 shows that the width and the length of the PH increase in the presence of the metallic mask. Furthermore, the position of the PH moves close to the shadow surface of the oblate spheroid. The maximal length of the PH is observed for index grading $g = 1$, corresponding to Fig. 2.8b. From Fig. 2.8 it is also followed that the curvature of the photonic hook depends on the index grading type parameter [31].

Figure 2.9 shows the key characteristics of the PH – FWHM and tilt angle as a function of the index grading type parameter g. One can see that controlling the parameter g makes it possible to adjust the PH curvature about 2 times – from 11.3° to 22.8°. Analysis [31] shows that both the large curvature of the PH and the maximum intensity could be obtained by the oblate spheroid at relative high g value. As follows from Fig. 2.8, the FWHM of the PH from the graded-index oblate spheroid is less than simple diffraction limit (less than $\lambda/2$) [32, 33] at $g > 1$ – minimal value of FWHM at maximal tilt angle corresponds to graded index $g \geq 1$. On the other hand, the minimum FWHM (about $\lambda/4$) of the PH is obtained at $g = 2$. One can also see that the tilt angle increases and the FWHM decreases as g-parameter increases.

It could be noted that to date, a compact and simple approach for generating electrically controlled Airy-like beams in high efficiency has remained challenging [21, 34, 35]. A polarization-controllable switch between dual Airy beams of orthogonal circular polarization was presented in [34] based on a polymer-dispersed liquid crystal. But this method has all the disadvantages of the methods for generating Airy-like beams [21, 35]. But in the case of the photonic hook, a change in the polarization of the incident radiation can be used to control the PH curvature [31]. Figure 2.10 illustrates the PH curvature at different polarization states of incident wave. This raises an intriguing property that the curvature of the PH changes from concave to convex by adjusting the incident electric field orientation from the y direction to z direction. Wherein the FWHMs of the PHs are of the same value (a little less than $\lambda/2$) for both polarizations, the PH curvatures are opposed. This property settles a fundamental requirement for PH applications, which has great potentials in the fields of biology science and optical manipulations, micro- and nano-fabrications, PH switching, and even some unchartered areas.

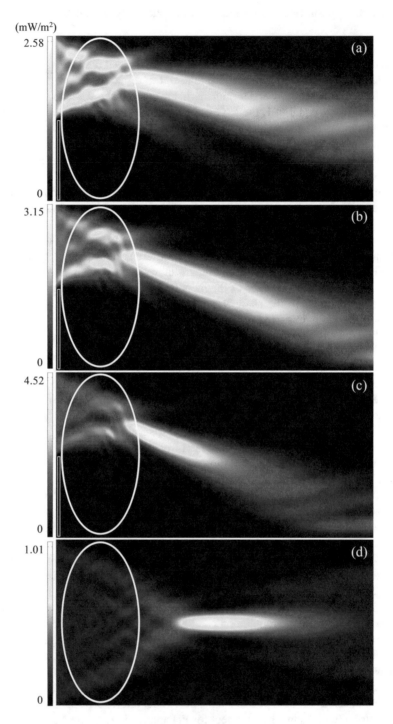

Fig. 2.8 Power flow (P = E × H) maps of the photonic hooks formed in the vicinity of graded-index oblate spheroid with mask at (**a**) $g = 0.2$ (concave), (**b**) $g = 1$ (linear), (**c**) $g = 2$ (convex), and (**d**) without mask

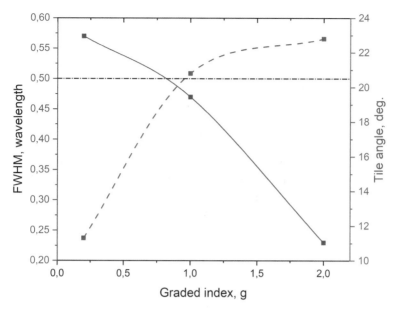

Fig. 2.9 FWHM and tilt angle as a function of the index grading type parameter g. The horizontal dash-dotted line corresponds to the simple diffraction limit of $\lambda/2$. Adapted from [31] with permission by IOP Publishing

Thus, the PH can be simply created using a mesoscale dielectric particle with asymmetric illumination, for example, by blocking beam by an amplitude mask. This method does not require the manufacture of microparticles with a special shape or complex internal structure. The mesoscale dimensions and simplicity of the PHs are much more controllable for practical tasks and indicating a wide range of potential applications. Finally, we would like to stress that the observed method of PH generation should be inherent to other beams including surface waves, microwaves, and acoustics for interacting with mesoscale symmetric obstacles and asymmetric illumination.

2.3   Steering and Bending of the Photonic Jet

The elucidation of PNJ formation mechanisms in the presence of a metal screen, which encompasses the dielectric object without gaps, was studied in [36]. It has been shown that the asymmetric control of the tangential component of the energy flux inside the mesoscale particle (both spherical and cubic) is possible by metal screen along or across one side of the particle. As example, in Fig. 2.11 the metal screen partially covers only the upper face of the cubic particle, and the power flow shifts toward the upper face. In the presence of a perfectly conducting screen, the field component at the place of its contact with the dielectric disappears, and a reflected wave appears. As a result of reflection, the amplitude of the tangential

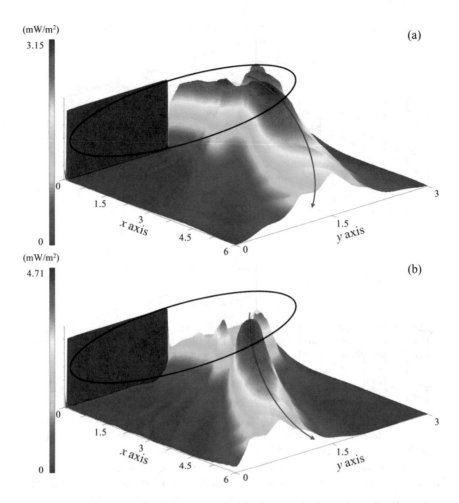

Fig. 2.10 Power flow patterns of the photonic hooks (left) and E-field distributions for incident electric field (right) formed by graded-index oblate spheroid with mask at $g = 1$ for incident electric field along the (**a**) z direction and (**b**) y direction. Reprinted from [31] with permission by IOP Publisher

electric field component increases in the direction opposite to that of wave propagation and weakens in the direction of wave propagation. At the same time, the distribution of the magnetic field component also changes, which leads to the formation of the power flow density components directed toward the axis of symmetry in front of the screen and to the displacement of the power flow density concentration area in the direction of wave propagation inside the object. From Fig. 2.11 one can see that control of the tangential electric field component by metal screen placed on one side of the surface of the particle allows deflecting and bending of the PNJ, which

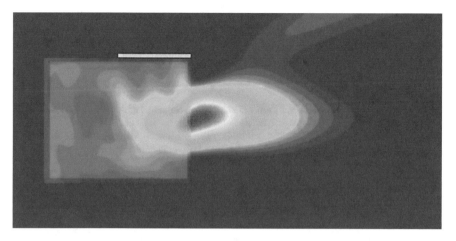

Fig. 2.11 Field intensity distributions when the metal screen partially covers only the upper face of the cubic particle

opens up a new possibility of the formation of specified configurations of localized fields. So we propose a mechanism to steer and tailor PNJ by using mask along with particle side concepts. But this is the task of future research.

References

1. F. Wang, L. Liu, P. Yu, Z. Liu, H. Yu, Y. Wang, and W. Li. Three-Dimensional Super-Resolution Morphology by Near-Field Assisted White-Light Interferometry // Sci. Rep. 6, 24703 (2016).
2. E. Xing, H. Gao, J. Rong, S. Khew, H. Liu, C. Tong, and M. Hong. Dynamically tunable multi-lobe laser generation via multifocal curved beam // Opt. Express 26, 30944 (2018).
3. J. Yang, P. Twardowski, P. Gérard, Y. Duo, J. Fontaine, and S. Lecler. Ultra-narrow photonic nanojets through a glass cuboid embedded in a dielectric cylinder // Opt. Express 26, 3723 (2018).
4. Y. Huang, Z. Zhen, Y. Shen, C. Min, and G. Veronis. Optimization of photonic nanojets generated by multilayer microcylinders with a genetic algorithm // Opt. Express 27, 1310 (2019).
5. L. Yue, B. Yan, J. Monks, Z. Wang, N. T. Tung, V. D. Lam, O. V. Minin, and I. V. Minin. Production of photonic nanojets by using pupil-masked 3D dielectric cuboid // J. Phys. D: Appl. Phys. 50, 175102 (2017)
6. L. Yue, B. Yan, J. Monks, Z. Wang, N. T. Tung, V. D. Lam, O. V. Minin, and I. V. Minin. A millimetre-wave cuboid solid immersion lens with intensity-enhanced amplitude mask apodization // Journal of Infrared, Millimeter, and Terahertz Waves, 39(6), 546–552 (2018)
7. L. Yue, B. Yan, J. Monks, Z. Wang, I. V. Minin, O. V. Minin. Intensity-enhanced apodization effect on an axially illuminated circular-column particle-lens // Ann. Phys. (Berlin), 1700384 (2017)
8. I. V. Minin, O. V. Minin, C.-Y. Liu, H.-D. Wei, Y. Geints, and A. Karabchevsky. Experimental demonstration of tunable photonic hook by partially illuminated dielectric microcylinder // Optics Letters, 45(17), 4899–4902 (2020)

9. L. Yue, O. V. Minin, Z. Wang, J. Monks, A. Snalin, and I. V. Minin. Photonic hook: a new curved light beam // Opt. Lett. 43, 771 (2018).
10. I. V. Minin, O. V. Minin, G. Katyba, N. Chernomyrdin, V. Kurlov, K. Zaytsev, L. Yue, Z. Wang, and D. Christodoulides. Experimental observation of a photonic hook // Appl. Phys. Lett. 114, 031105 (2019).
11. Y. Geints, I. V. Minin, and O. V. Minin. Tailoring 'photonic hook' from Janus dielectric micro-bar // J. Opt. 22, 065606 (2020).
12. A. Ang, A. Karabchevsky, I. V. Minin, O. V. Minin, S. Sukhov, and A. Snalin, Photonic Hook based optomechanical nanoparticle manipulator // Sci. Rep. **8**, 2029 (2018).
13. I. V. Minin, O. V. Minin, C.-Y. Liu, and H.-D. Wei, A. Karabchevsky. Simulation and experimental observation of tunable photonic nanojet and photonic hook upon asymmetric illumination of a mesoscale cylinder with mask // ArXiv: 2004.05911 (2020)
14. Z. Cao, C. Zhai, J. Li, F. Xian, and S. Pei, Light sheet based on one-dimensional Airy beam generated by single cylindrical lens, // Opt. Commun. 393, 11–16 (2017).
15. M. Avendaño-Alejo, L. Castañeda, and I. Moreno, Properties of caustics produced by a positive lens: meridional rays, // J. Opt. Soc. Am. A **27**, 2252–2260 (2010).
16. N. Yu, P. Genevet, M. Kats, F. Aieta, J. Tetienne, F. Capasso, and Z. Gaburro. Light Propagation with Phase Discontinuities: Generalized Laws of Reflection and Refraction // Science 334, 333 (2011).
17. H. Yang, R. Trouillon, G. Huszka, and M. Gijs. Super-Resolution Imaging of a Dielectric Microsphere Is Governed by the Waist of Its Photonic Nanojet // Nano Lett. 16, 4862 (2016).
18. C. Liu and W. Lo. Large-area super-resolution optical imaging by using core-shell microfibers// Opt. Commun. 399, 104 (2017).
19. P. Ferrand, J. Wenger, A. Devilez, M. Pianta, B. Stout, N. Bonod, E. Popov, and H. Rigneault. Direct imaging of photonic nanojets// Opt. Express 16, 6930 (2008).
20. B. Luk'yanchuk, R. Paniagua-Domínguez, I. V. Minin, O.V. Minin, Z. Wang, Refractive index less than two: Photonic nanojets yesterday, today and tomorrow (Invited). // Opt. Mater. Express, 7, 1820–1847 (2017).
21. I.V. Minin, C.-Y. Liu, Y. E Geints, O. V. Minin. Recent advantages in Integrated Photonic Jet-Based Photonics. // Photonics, 7(2), 41 (2020)
22. I. V. Minin and O. V. Minin. Recent Trends in Optical Manipulation Inspired by Mesoscale Photonics and Diffraction Optics // J of Biomedical Photonics & Eng 6(2), 020301 (2020)
23. S. Kong, A. Taflove and V. Backman, Quasi one-dimensional light beam generated by a graded-index microsphere, // Opt. Express, 17, 3722 (2009)
24. Y. Geints, A. Zemlyanov and E. Panina, Photonic nanojet calculations in layered radially inhomogeneous micrometer-sized spherical particles // J. Opt. Soc. Am. B, 28, 1825 (2011)
25. Y. Shen, L. Wang and J. Shen, Ultralong photonic nanojet formed by a two-layer dielectric microsphere // Opt. Lett., 39, 4120 (2014)
26. G. Gu, R. Zhou, Z. Chen, H. Xu, G. Cai, Z. Cai and M. Hong, Super-long photonic nanojet generated from liquid-filled hollow microcylinder // Opt. Lett., 40, 625 (2015)
27. C. Liu, T. Yen, O. V. Minin and I. V. Minin, Engineering photonic nanojet by a graded-index micro-cuboid // Physica E, 98, 105 (2018)
28. Z. Zhen, Y. Huang, Y. Feng, Y. Shen and Z. Li, An ultranarrow photonic nanojet formed by an engineered two-layer microcylinder of high refractive-index materials // Opt. Express, 27, 9178 (2019)
29. Y. Huang, Z. Zhen, Y. Shen, C. Min and G. Veronis, Optimization of photonic nanojets generated by multilayer microcylinders with a genetic algorithm // Opt. Express, 27, 1310 (2019)

30. Y. Wang, C. Dai and J. Li, Numerical study of tunable photonic nanojets generated by biocompatible hydrogel core-shell microspheres for surface-enhanced Raman scattering applications // Polymers, 11, 431 (2019)

31. C.-Y. Liu, H.-J. Chung, O. V. Minin and I. V. Minin. Shaping photonic hook via well-controlled illumination of finite-size graded-index micro-ellipsoid// J. of Optics 22(8), 085002 (2020)

32. E. Abbe, Beiträge zur Theorie des Mikroskops und der mikroskopischen Wahrnehmung // Archiv für Mikroskopische Anatomie, 9, 413–468 (1873); https://doi.org/10.1007/BF02956173

33. Lord Rayleigh F.R.S. XXXI. Investigations in optics, with special reference to the spectroscope // The London, Edinburgh, and Dublin Philosophical Magazine and Journal of Science, 8(49), 261–274 (1879); doi: https://doi.org/10.1080/14786447908639684

34. B. Wei, P. Chen, W. Hu, W. Ji, L.-Y. Zheng, S-J. Ge, Y. Ming, V.Chigrilov and Y.-Q. Lu, Polarization-controllable Airy beams generated via a photoaligned director-variant liquid crystal mask // Sci Rep 5, 17484 (2015)

35. N. Efremidis, Z. Chen, M. Segev, D. N. Christodoulides, Airy beams and accelerating waves: An overview of recent advances // Optica 6, 686–701 (2019).

36. I. Dorofeev, V. Suslyaev, I. V. Minin, O. V. Minin. The role of the tangential electric field component to the terahertz jet and hook formation by dielectric cube and sphere // Opt Eng 60(8), 082003 (2020)

Chapter 3
Acoustic Hook

Abstract In 2017, it has been demonstrated for the first time that the existence of an acoustic analogue of photonic jet phenomenon, called acoustojet, providing for subwavelength localization of acoustic field in shadow area of an arbitrary 3D penetrable mesoscale particle, is possible. In this chapter, we report the first experimental observation (2020) of a new type of near-field curved acoustic beam – acoustic hook – confirmed by simulations. This new curved acoustical beam is generated by asymmetric distribution of the vortices in a polymer Janus particle immersed in water. The origin of the vortices is in the conversion of an incident longitudinal wave mode to a shear wave in a solid and then back to a longitudinal wave in the water and has unique features, the radius of curvature of acoustic hook is less than the wavelength. Indeed, it is the smallest radius of curvature ever recorded for any acoustical beam. These results may be potentially useful when an object, located in the path of the beam, must be avoided and also for generation of an acoustical bottle beam.

3.1 Introduction

The characteristics of Airy-family beams are of great interests to the acoustic community [1]. For instance, designed Airy-like acoustic beams may bypass a target using their bending feature. Acoustic self-bending waves have been reported in [2, 3]. Airy-like sound waves were demonstrated in [4], and accelerating underwater acoustic waves were discussed in [5]. Surface gravity water waves, both linear and nonlinear, were observed in [6]. For the generation of an acoustic Airy-like beam without using active multi-element acoustic sources, grooves on the surface of a piston acoustic transducer (corrugated transducer) have been discussed in [7]. Formation of an acoustic Airy beam by dynamic diffractive optical phase element [8] or metasurface [9] was also considered. However, Airy acoustic beams have all the disadvantages of optical Airy-like beams mentioned in the Introduction to the book. Below we will briefly consider the features of the formation and the main properties of acoustic hooks, based on the phenomenon of an acoustic jet, first proposed in [10] and named acoustojet.

O. V. Minin, I. V. Minin, *The Photonic Hook*, SpringerBriefs in Physics,
https://doi.org/10.1007/978-3-030-66945-4_3

3.2 Acoustic Jet

Subwavelength focusing methods in acoustics, as a rule, require sophisticated sig-
nal processing methods or are based on new artificial materials, which limit practical applications. At the same time in [11] a problem of plane acoustic wave focusing
by spherical gas-filled lens with Mie size parameter of $2\pi R/\lambda = (17.5 \dots 27.5)$ was
discussed. It was noted that with small Mie parameters, the location of focus near
the shadow surface of such lens did not correspond to focal distance, prognosticated
by geometrical optics law. On the other hand, it should be expected that methods of
subwavelength focusing, based on optical PNJ phenomenon, can be successfully
used in the acoustic range in the linear mode as well. This can be stated based on the
formal analogy between the acoustic and electromagnetic wave equations [12–14].
Pioneering research of scalar acoustic wave scattering at penetrable sphere shown
[10, 14] that for acoustic wave a formation of acoustic analogue of photonic jet with
width of a jet beyond the diffraction limit (as small as 0.28 of wavelength) for different particle materials (plastic, lead, silver, droplet, etc.) is possible. However, the
main difficulty in application of the acoustic-electromagnetic analogy to a PNJ phenomenon is the different physical nature of transverse vectorial electromagnetic
field and scalar longitudinal acoustic field used in 3D configuration [15]. This
causes certain complexity in establishing direct link between the measurable parameters such as acoustic pressure and electromagnetic field intensity. As it is followed
from optics [16, 17], the lens material should have a bigger refractive index than the
surrounding medium, which means that the speed of light in the medium should be
higher than in lens material:

$$n_{lens} / n_{medium} = m, \rightarrow v_{lens} / v_{medium} = 1 / m,$$

where n is the refractive index, v is the speed of light in the medium, and m is an
optimal value of effective refractive index contrast. But such a simplified relationship between the refractive index contrast in optics and the ratio of the speeds of
sound in acoustics, used, for example, in [18], is valid only approximately. For
example, an acoustical lens based on a thin hollow ABS spherical container structure, filled with different compatible liquids, produces a dynamically shaped acoustojet by either shifting the operating frequency or modifying the geometry of the
lens [19]. And it was noted that the density of the material of the particle lens also
affects its focusing properties, and, therefore, this parameter, which has no analogue
in the electromagnetic range, should be taken into account in acoustics. It is due to
the propagation of acoustic waves are controlled by the bulk modulus of a material
and density of a material:

$$\nabla^2 P - \frac{\rho}{B} \frac{\partial^2 P}{\partial t^2} = 0,$$

where P is the pressure and B and ρ are the bulk modulus of materials and mass density, respectively. These two main parameters are analogous to permittivity ε and permeability μ of the electromagnetic parameters. Moreover, in solid mesoscale particles, the effect of the transverse (shear) velocity [20] on the formation of an acoustic jet is significant – while in optics most dielectrics are isotropic, in acoustics solids are anisotropic. Several works presented modification of the acoustojet properties by changing the mechanical properties of the lens, by modifying the relative Mie size parameters of the lens (both natural or artificial) [21–29]. In [26] it was discovered that sonic composite lens, which is vertically extended from a 2D flat phononic crystal (PhC) structure (Fig. 3.1), is able to focus acoustic waves in a three-dimensional manner in the form of acoustojet. It was shown that the transverse beamwidth (FWHM) can be smaller than the fundamental diffraction limit (~0.3λ) (Fig. 3.2).

It could be noted that from Fig. 3.2b, it is followed that near the shadow surface of cubic particle, a forklike region of low intensity (similar to Bessel vortex beams with an axial null [30]) where particles may be trapped in application to acoustical tweezer is formed.

Also recently, it was discovered that when a high-numerical aperture acoustical zone plates [31–35] or acoustical particle lens is selectively blocked by using a pupil mask, it will produce some novel focusing properties [36–38], for example, increasing the sound intensity and the resolution due to the local abnormal apodization effect, for the first time discovered in optics [39–41]. These research show that by selecting the ratio of the lens and surrounding medium materials, as well as the lens Mie size parameter, it is possible to localize acoustic energy to a spot smaller than $\lambda/2$ in non-resonant mode. Moreover, a giant field enhancement mode caused by microsphere's internal partial waves and which were define as a "super-resonance" effect was discovered [29]. The enhancement factor can be extremely large at these "hot spots," reaching the order of 10^4–10^7 (see Fig. 3.3). It could be noted that for the artificial materials (metamaterials), one generally assumes that its electromagnetic

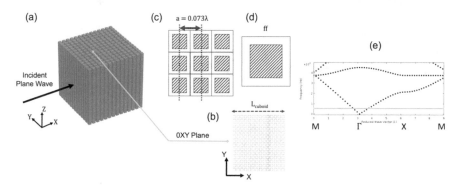

Fig. 3.1 (**a**) 3D scheme of the metamaterial cuboid structure the plane wave travels along X direction, (**b**) plane XY view of the crystal, (**c**) zoom of XY plane where a is the lattice constant and (**d**) the filling fraction is the relation between the area occupied by the material and the area of the unit cell. (**e**) Band structure of the unit cell along M-Γ-X-M. Reprinted under CC BY from [26]

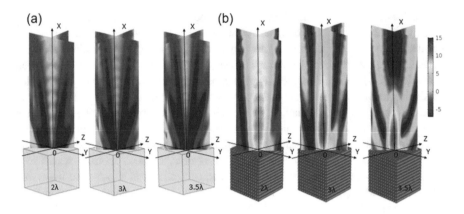

Fig. 3.2 Gain pressure level distribution for different cuboid size and nature (**a**) rigid cuboid, (**b**) metamaterial 3D cuboid with 2D internal PhC structure. The plane wave travels along X axes from bottom to top in the figures. Reprinted under CC BY from [26]

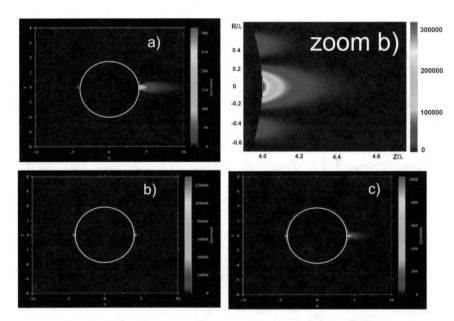

Fig. 3.3 Resonant scattering on Rexolite sphere with radius of R = 4λ (Mie size parameter q = 25.132766) immersed in water at frequency of 1 MHz: relative density contrast is 1.0402, and the speed of sound contrast is 1.570. At the super-resonance condition, the enhancement factor can be extremely large at "hot spots" (**b**), reaching the order of 10^4–10^7. It could be noted that at the super-resonance condition, the hot spot has a super resolution about of 0.21–0.23λ, which exceeds solid immersion resolution limit. Figures (**a**) and (**c**) demonstrate the formation of an acoustic jet before and after resonance, respectively. Modified from [29] with the permission of the Acoustical Society of America

(acoustic) response coincides with that of a homogeneous medium, and suitable phenomenological material parameters (such as effective permittivity and/or permeability, sound, and shear velocity) can be introduced for describing the effective medium response. But in the case of mesoscale object, the concept of an effective environment is not defined, and the use of metamaterials is not obvious. Some of practical realization and applications of the acoustojet effect was also discussed [42–46].

3.3 Acoustic Hook

A flat polymer (PMMA and cross-linking polystyrene with divinyl benzene, Rexolite [41]) wavelength-scaled cuboid particle sound lens [27] and spherical Rexolite lens [21] immersed in water and which can form an acoustojet with no need of negative refraction were studied. The choice of lens material was dictated by the acoustic impedance and the existence of shear waves. Rexolite cuboid lens was determined by its shear sound speed ($c_{s\text{Rexolite}} = 1157$ m s^{-1}), longitudinal sound speed ($c_{l\text{Rexolite}} = 2337$ m s^{-1}), and density ($\rho_{\text{Rexolite}} = 1049$ kg m^{-3}) [27, 47]. The surrounding medium was water: $c_{\text{water}} = 1500$ m s^{-1} and $\rho_{\text{water}} = 1000$ kg m^{-3}. Experimental investigation of the focusing properties of the cuboid lens was made using the ultrasonic immersion transmission technique with a total precision automated measurement system [27] with a fixed-piston ultrasonic transducer used as an emitter at a central frequency of 250 kHz. As a receiver, a polyvinylidene fluoride needle hydrophone with a diameter of 1.5 mm was used (Fig. 3.4). A detailed description of the experiments can be found in [27].

Simulations have shown that the FWHM in the case of two wavelength dimension cuboid is 0.44λ, which is less than the diffraction limit [48, 49]. For the case of 2.5 wavelength cube dimensions, FWHM is about 0.25λ. Similar to optical case, the mechanism of PNJ formation is due to two main effects (Fig. 3.5). First, a wave inside a cubic particle in the center of the cube propagates with a lower phase veloc-

(a)   **(b)**

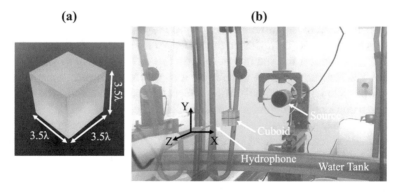

Fig. 3.4 Dimensions of (**a**) Rexolite cuboid particle lens measured and (**b**) experimental setup. Reprinted from [27] with the permission from Elsevier

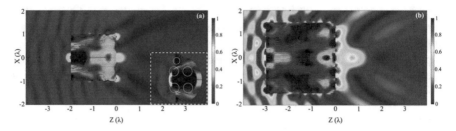

Fig. 3.5 Normalized relative intensity flow in XZ planes for cuboid particle lens of different sizes: (**a**) 2λ (the inset corresponds to an enlargement of the illuminated face of the cuboid. Vortices are indicated in the circles) and (**b**) 3λ. Reprinted from [27] with the permission from Elsevier

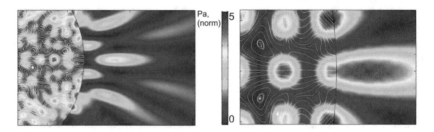

Fig. 3.6 Mechanical energy flux (Pa) near the shadow surfaces of the Rexolite (left) and ABS (right) spheres (not in scale), immersed in water. The vortices distributions are clearly visible

ity than that near the edge of the cube. Second, a curvature of the normally incident plane wavefront at the edge of the cuboid is such that the wave is directed into the cuboid from the edges to the center, forming a focus in the form of acoustojet. Analysis of the flow diagrams inside the particles has shown (Fig. 3.6) that there are regions where the intensity lines produce vortices [50–52], in such a way the vortices near the illuminated side of the particle redirect the intensity flow to different areas of the lens. In the case of PMMA material, the vortices are not found near the cuboid, resulting in a different redistribution of the energy inside the cuboid for Rexolite [57] and PMMA.

This fact is due to the different sound speed contrast of the materials and surrounding medium and more lower ratio of c_l/c_s for PMMA [27]. Experimental investigations confirm the simulation results (Fig. 3.7). It has been shown that at particle sizes comparable to the wavelength it is possible to form an acoustojet with a cuboid particle lens due to diffraction effects that could overcome the refractive effects of a conventional cylindrical or spherical particle lens.

Based on the results of studies of a cubic particle lens, the corresponding studies were carried out for a particle of a cubic shape with a broken symmetry to compare the features of the formation of a photonic [53–55] and an acoustic [56] hooks. As it was noted above, in optics three key parameters are decisive: particle 3D shape, particle Mie size parameter, and refractive index contrast [16]. In acoustics, unlike optical light due to the different nature of the waves, additional two parameters are

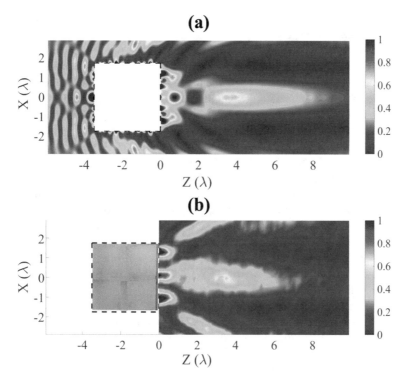

Fig. 3.7 Comparison of (**a**) simulated and (**b**) measured normalized sound pressure distribution in XZ planes for cuboid particle of 3.5λ size. Reprinted from [27] with the permission from Elsevier

determining – total five parameters at least: 3D shape of the particle, the Mie parameter of the particle, the contrast of the speed of sound in the medium and the particle, and the contrast of the density of the material of the medium and particle. In addition for solid particle lens, the transverse (shear) speed of sound is important. This fact causes the mesoscale particles to become anisotropic. Therefore, the possibility of implementing the effect of the acoustic hook (AH) was not so obvious [56].

Simulations have shown that in the case of Janus particle consisting of Rexolite cuboid with a rectangular trapezoid having an interior angle of $\alpha = 10°$, the shape of the acoustojet is slightly bent. As the interior angle-α increases up to 20°, the curvature of the acoustojet increases. At the same time, at the interior angle-$\alpha > 20°$, the effect of the acoustojet curvature weakens. In analogy with Refs.[54, 55], the AH curvature is determined by an angle-β between the two lines, which link the end point with inflection point and the inflection point with the start point, respectively; see Fig. 3.9. The results show that as the trapezoid angle-α increases, the angle of curvature-β increases due to $\alpha < 30$ for cuboid with 3 wavelength in dimensions (see Fig. 3.10). As in optics [54, 55], in the case of AH, only the main lobe has a curved shape in contrast to the family of curved sidelobes for Airy-family beams. It is notable the pressure maximum of AH not located on the shadow surface of the Janus particle and shifted to a distance of about 0.17λ from the surface of the particle, in contrast to optical case, where this maximum is located at the rear surface of the particle. In the case of a Janus particle, minimal FWHMs of acoustic hook have subwavelength values and are about 0.72λ and 0.83λ at the shadow surface of the Janus particle and at the point of maximal pressure along AH, respectively.

To experimentally verify the phenomenon of AH formation and the self-bending effect, the ultrasonic immersion technique was used, and a rectangular trapezoidal particle was machined from Rexolite material with the $l = 3\lambda$ (in water) side cuboid and with an interior angle of $\alpha = 20°$ (Fig. 3.8).

Measured sound pressure distribution is presented in Fig. 3.9, which clearly shows the effect of a curvilinear region formation of the acoustic field localization behind a dielectric Janus particle immersed in water. One can see that an experimental curvature $\beta = 144°$ is obtained which agree well with the simulated results shown in Fig. 3.10 (note that the angle-$\beta = 180$ corresponds to a non-curved acoustojet). The curvature characterization and pressure distributions along the acoustic hook are presented in Table 3.1. One can see that the width (FWHM/λ) of the acoustic hook has subwavelength value. Moreover, as it follows from the Table 3.1 (see the cross-section III), for the Janus particle FWHM/$\lambda = 0.56$, but for the cuboid particle this value is FWHM/$\lambda = 1.17$.

To explain the AH phenomenon, let us consider the relative intensity ($\vec{I} = p\hat{A}\cdot\vec{u}$, where p is the acoustic pressure and $\vec{u}$ is the particle velocity) flow diagrams, shown in Fig. 3.5. From Fig. 3.11a one can see that there are regions inside the regular cuboid where the intensity lines produce vortices that in its turn redirect the intensity flow to different localized areas within the cuboid [27, 56]. The distribution of these vortices is symmetrical due to the regular cuboid's symmetry, and so the acoustojet near the shadow surface of the particle is symmetrical too.

However, in the case of the cuboid particle with broken symmetry (Fig. 3.11b), one can see that the distribution symmetry of the vortices inside the particle is also broken. Such a distribution produces the curvature of the AH. The nature of the vortices is in the conversion of an incident longitudinal wave mode to transverse (shear) wave in a solid and then back to a longitudinal wave in the water [56]. The fields of the particle velocity $\vec{v}$ can be described by gradients of the scalar function φ in a fluid [56] ($\vec{v} = \vec{\Delta}\varphi$) and the curl of a vector field $\vec{\Psi}$ in a solid $\vec{v} = \vec{\nabla}\varphi + \vec{\nabla} \times \vec{\Psi}$, where $\vec{\nabla} \times \vec{\Psi}$ and $\vec{\nabla}\varphi$ represent the shear and the longitudinal waves, respectively. The vortices in the intensity flow in a solid are due to transverse waves, i.e., $\vec{\nabla} \times \vec{\Psi} \neq 0$.

Fig. 3.8 Janus Rexolite particle lens, immersed in water. λ – the wavelength in water. Reprinted from [56] with the permission from Elsevier

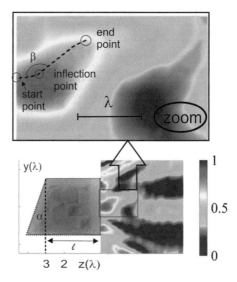

Fig. 3.9 Measured sound pressure distribution for experimental confirmation of an acoustic hook phenomenon and definitions of curvature. Reprinted from [56] with the permission from Elsevier

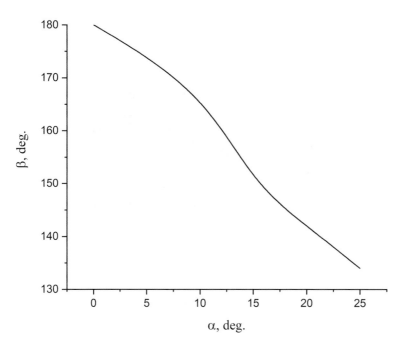

Fig. 3.10 Simulation results of the curvature-β (see definitions in Fig. 3.9 above) of an acoustic hook vs trapezoid angle-α for 3λ Janus Rexolite particle

Table 3.1 Experimental values of an acoustic hook parameters

#Cross-sections	X_{max}/λ	Z/λ	P/P_{max}	FWHM/λ
I	0.08	0.33	0.92	0.72
II	−0.08	0.33	0.90	0.82
III	−0.08	0.50	0.79	0.82
IV	−0.25	0.50	0.79	0.83
V	−0.25	0.67	0.72	0.56
VI	−0.42	0.67	0.71	0.58
VII	−0.42	0.83	0.69	0.71
VIII	−0.42	1.00	0.58	1.06

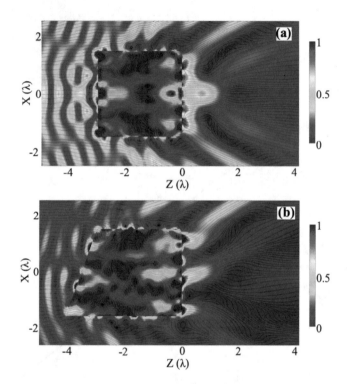

Fig. 3.11 Simulation results normalized relative intensity flow for (**a**) 3λ side regular cuboid and (**b**) Janus particle with an interior angle of α = 20°. Reprinted from [56] with the permission from Elsevier

The constructive interference of the shear and longitudinal waves creates specific localized pressure distribution areas inside the particle and near its shadow surface, which is a kind of boundary conditions for the wave propagating through the surrounding medium. It means that if we completely exclude the particle lens from consideration, and we take the pressure distribution on its shadow surface as the boundary conditions of an "effective" source of radiation, then the effect of the acoustojet or acoustic hook formation will be also observed.

Thus one can see that generation of the subwavelength curved acoustical beams is possible due to interaction of an acoustic wave with Janus dielectric particle with broken symmetry in shape. One of the unique properties of the acoustic hook, similar to optical case, is the radius of the beam curvature which is substantially smaller than the wavelength in water. Therefore, it has been shown that the concept of electromagnetic photonic hook has gone beyond optics [53–55] and now penetrated acoustics [56].

3.4 Acoustic Hook Generated by Cylindrical Particle with Off-Axis Pupil Mask

In [58] the influence of the beam waist on optical "photonic nanojet" (PNJ) parameters (width, length, and intensity) was analyzed. It was concluded that under a spatially limited light beam illuminating a transparent spherical particle, main PNJ parameters are optimal when the beam waist radius of the illuminating light is comparable with the particle radius. When the spatially limited beam shifting from the axis of symmetry to the particle edge the PNJ becomes much shorter with decreasing of its intensity, and photonic jet bends toward the axis of symmetry of the particle, the curvature of the photonic jet was not observed.

In Chap. 2 it has been shown [59] that the optical hook could be simply created using a symmetric mesoscale dielectric particle with asymmetric illumination, for example, by blocking illuminated beam by an amplitude mask. It was noted that this method should be inherent to other beams including acoustics for interacting with mesoscale symmetric penetrating obstacles under structured asymmetric illumination.

The use of an amplitude mask to control the properties of an acoustic jet was considered in [37]. Cylindrical lens consists of an inner core with the radius $R_{in}/\lambda = 3.29$ mm and an outer radius $R_{out}/\lambda = 3.33$. As a material of the shell of cylindrical lens (filled with a liquid) and mask container, acrylonitrile butadiene styrene (ABS) was selected [60, 61]. The ABS material parameters are as follows: compression wave velocity ($c_p = 2250$ m/s), shear wave velocity ($c_s = 1447$ m/s), and density ($\rho_s = 1050$ kg/m^3). The whole structure was immersed in water. The exposed and covered masks were placed on the front surface of the cylindrical particle lens. The rest of the simulation details can be found in [37].

Figure 3.12a shows the configuration of a cylindrical particle with a mask, similar to that studied in optics [59]. It can be seen that, similar to the electromagnetic range, a curvilinear localized pressure area is also formed in acoustics near the shadow part of the cylinder. On the other hand, if we restrict the acoustic field incident on the particle from the side of the free upper surface of the particle (Fig. 3.12b), then such structured irradiation makes it possible to control both the curvature and the length of the formed acoustic hook. The nature of the change in the shape of the acoustic beam in the second variant of the off-axis mask consists in the partial blocking of the tangential component of the flow along the upper part of the particle,

Fig. 3.12 Formation of an acoustic hook behind a cylindrical mesoscale particle-lens during structured irradiation using masks of different configurations

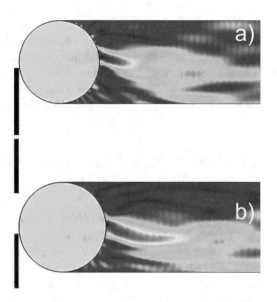

which leads to additional curvature of the formed beam at the exit of the particle [62]. Thus, spatially limited acoustic beam allows controlling three-dimensional shape of the acoustic hook.

Conclusion The detailed pioneering research in acoustics show that the origin of the acoustojet formation and bent phenomena is the vortices that appear inside the mesoscale solid particles due to the conversion of the incident longitudinal wave mode to a shear wave in a solid. Both the acoustojet and acoustic hook have unique features of subwavelength beamwidth, and AH are self-bending with radius of curvature that is substantially smaller than the wavelength and represents the smallest radius of curvature ever recorded for any acoustical beams including comparison of Airy-like acoustical family beams. Such properties provide the impetus to further develop improved subwavelength microscopy and acoustical tweezer for various applications especially in on-chip configuration.

One can notice that solid mesoscale particle may be manufacturing from an elastic flexible material, and the resultant acoustojet and/or acoustic hook can be adjusted by method of mechanical compression or stretch.

Therefore, it has been shown that the concept of electromagnetic photonic hook has gone beyond optics [53–55] and now penetrated acoustics [56].

References

1. N. Efremidis, Z. Chen, M. Segev, D. N. Christodoulides. Airy beams and accelerating waves: An overview of recent advances // Optica 6, 686–701 (2019).

2. P. Zhang, T. Li, J. Zhu, X. Zhu, S. Yang, Y. Wang, X. Yin, and X. Zhang, Generation of acoustic self-bending and bottle beams by phase engineering // Nat. Commun. 5, 4316 (2014).
3. K. Mohanty, S. Mahajan, G. Pinton, M. Muller, Y. Jing. Observation of Self-Bending and Focused Ultrasound Beams in the Megahertz Range // IEEE Trans. on Ultrasonics, Ferroelectrics, and Frequency Control, 65(8), 1460–1467 (2018)
4. S. Zhao, Y. Hu, J. Lu, X. Qiu, J. Cheng, and I. Burnett, Delivering sound energy along an arbitrary convex trajectory // Sci. Rep. 4, 6628 (2014).
5. U. Bar-Ziv, A. Postan, and M. Segev, Observation of shape-preserving accelerating underwater acoustic beams // Phys. Rev. B 92, 100301 (2015).
6. S. Fu, Y. Tsur, J. Zhou, L. Shemer, and A. Arie, Propagation dynamics of Airy water-wave pulses // Phys. Rev. Lett. 115, 034501 (2015).
7. Z. Lin, X. Guo, J. Tu, Q. Ma, J. Wu, and D. Zhang. Acoustic non-diffracting Airy beam // Journal of Applied Physics 117, 104503 (2015)
8. A. Benstiti, K. Ferria, and A. Bencheikh, Generation of a variety of Airy beams using a dynamic diffractive optical phase element // J. Opt. Soc. Am. B 37, A45–A53 (2020)
9. D.-C. Chen, X.-F. Zhu, D.-J. Wu, and X.-J. Liu. Broadband Airy-like beams by coded acoustic metasurfaces // Appl. Phys. Lett. 114, 053504 (2019)
10. I.V. Minin and O.V. Minin, Acoustojet: acoustic analogue of photonic jet phenomenon // ArXiv:1604.08146 (2016)
11. C. Thomas, K. L. Gee, and R. S. Turley. A balloon lens: Acoustic scattering from a penetrable sphere // American Journal of Physics 77, 197–203 (2009)
12. Y. Feng, J. Ge, and F. Wan, The derivation of scaling relationship between acoustic and electromagnetic scattering by spheres // AIP Advances 3, 112130–9 (2013).
13. L. Nicolas, M. Furstoss and M. A. Galland, Analogy electromagnetism-acoustics: Validation and application to local impedance active control for sound absorption // The European Physical Journal Applied Physics, 4, 95–100 (1998).
14. O. V. Minin and I. V. Minin. Acoustojet: acoustic analogue of photonic jet phenomenon based on penetrable 3D particle // Opt Quant Electron 49, 54 (2017)
15. D. Colton, R. Kress, *Inverse Acoustic and Electromagnetic Scattering Theory*, 388 p. Springer, Berlin (2013)
16. B. Luk'yanchuk, R. Paniagua-Domínguez, I. V. Minin, O.V. Minin, Z. Wang, Refractive index less than two: Photonic nanojets yesterday, today and tomorrow (Invited) // Opt. Mater. Express 7, 1820–1847 (2017).
17. I.V. Minin, C.-Y. Liu, Y. Geints, O. V. Minin. Recent advantages in Integrated Photonic Jet-Based Photonics. // Photonics 7(2), 41 (2020)
18. D. Veira Canle, T. Kekkonen, J. Mäkinen, T. Puranen, H. J. Nieminen, A. Kuronen, S. Franssila, T. Kotiaho, A. Salmi & E. Hæggström. Practical realization of a sub-λ/2 acoustic jet. // Sci Rep 9, 5189 (2019)
19. S. Perez-Lopez, P. Candelas, J. M. Fuster, C. Rubio, O. V. Minin, and I. V. Minin, Liquid-liquid core-shell configurable mesoscale spherical acoustic lens with focusing below the wavelength // Appl. Phys. Express 12, 087001 (2019).
20. Poisson, S. D. *"Mémoire sur la propagation du mouvement dans les milieux élastiques"* [Memoir on the propagation of motion in elastic media]. // Mémoires de l'Académie des Sciences de l'Institut de France (in French). 10, 549–605 (1831).
21. J. H. Lopes, M. A. B. Andrade, J. P. Leao-Neto, J. C. Adamowski, I. V. Minin, and G. T. Silva, Focusing acoustic beams with a ball-shaped lens beyond the diffraction limit // Phys. Rev. Appl. 8, 024013 (2017).
22. I.V. Minin and O. V. Minin. Mesoscale Acoustical Cylindrical Superlens // MATEC Web of Conferences 155, 01029 (2018)
23. C. Rubio, D. Tarrazó-Serrano, O. V. Minin, A. Uris, and I. V. Minin, Sound focusing of a wavelength-scale gas-filled flat lens // Europhys. Lett, 123, 64002 (2018).
24. S. Pérez-López, J. M. Fuster, I. V. Minin, O. V. Minin, and P. Candelas, Tunable subwavelength ultrasound focusing in spherical lenses using liquid mixtures // Sci. Rep. 9, 1 (2019).

25. C. Rubio, D. Tarrazó-Serrano, O. V. Minin, A. Uris, and I. V. Minin, Wavelength scaled gas-filled cuboid acoustic lens with diffraction limited focusing // Results Phys. 12, 1905 (2019).

26. S. Castiñeira-Ibáñez, D. Tarrazó-Ibáñez, P. Candelas, O. V. Minin, C. Rubio and I.V. Minin. 3D sound wave focusing by 2D internal periodic structure of 3D external cuboid shape // Results in Physics 15 (2019) 102582

27. D. Tarrazó-Serrano, C. Rubio, O. V. Minin, A. Uris, and I. V. Minin, Ultrasonic focusing with mesoscale polymer cuboid // Ultrasonics 106, 106143 (2020)

28. O. V. Minin and I. V. Minin. Extreme effects in field localization of acoustic wave: super-resonances in dielectric mesoscale sphere immersed in water // IOP Conf. Series: Materials Science and Engineering 516, 012042 (2019)

29. I.V. Minin, O.V. Minin, C. Rubio, A. Uris. Super-resonances in dielectric mesoscale sphere immersed in water: effects in extreme field localization of acoustic wave // Proc. Mtgs. Acoust. 38, 070001 (2019)

30. N.Jiménez, V.J.Sánchez-Morcillo, R.Picó, L.M.Garcia-Raffi, V.Romero-Garcia and K.Staliunas. High-order Acoustic Bessel Beam Generation by Spiral Gratings // Physics Procedia, 70, 245–248 (2015).

31. D. Tarrazo-Serrano, S. Castieira-Ibanez, O.V. Minin, C. Rubio, P. Candelas and I. V. Minin. Design of acoustical Bessel-like beam formation by tunable angular spectrum in Soret zone plate lens // Sensors, *19*(2), 378 (2019)

32. C. Rubio, D. Tarrazó-Serrano, S. Castiñeira-Ibáñez, O.V. Minin, P. Candelas, I.V. Minin. Reference radius in Fresnel Zone Plates to control ultrasound beamforming // Proc. Mtgs. Acoust. 38, 030008 (2019)

33. D. Tarrazo-Serrano, C. Rubio, O. V. Minin, P. Candelas, I. V. Minin. Manipulation of focal patterns in acoustic Soret type ZP lens by using reference radius/phase effect // Ultrasonics 91, 237–241 (2019)

34. D. Tarrazo-Serrano, S. Castieira-Ibanez, O. V. Minin, C. Rubio, and I. V. Minin. Tunable depth of focus of acoustical pupil masked Soret Zone Plate // Sensors & Actuators: A. Physical 286, 183–187 (2019)

35. D. Sukhanov, I. V. Minin, O. V. Minin, I. Kuzmenko, T. Muksunov, E. Sivkov, F. Emelyanov. Control of levitating particle in ultrasound field // MATEC Web of Conferences 155, 01017 (2018)

36. C. Rubio, D. Tarrazó-Serrano, O. V. Minin, A. Uris, and I. V. Minin, Enhancement of pupil-masked wavelength-scale gas-filled flat acoustic lens based on anomaly apodization effect // Phys. Lett. A 383, 396 (2019).

37. C. Lu, R. Yu, K. Wang, J. Wang, and D. Wu. Tunable acoustic jet generated by a masked cylindrical lens // Applied Physics Express, 13, 097003 (2020)

38. C. Rubio, D. Tarrazó-Serrano, O. V. Minin, A. Uris, and I. V. Minin, Sound focusing capability of a CO_2 gas-filled cuboid // Physics of wave phenomena 28(4), 333–337 (2020)

39. I.V. Minin, C.-Y. Liu, Y. E Geints, O. V. Minin. Recent advantages in Integrated Photonic Jet-Based Photonics // Photonics *7*(2), 41 (2020)

40. L. Yue, B. Yan, J. Monks, Z. Wang, N. T. Tung, V. D. Lam, O. V. Minin, and I. V. Minin. Production of photonic nanojets by using pupil-masked 3D dielectric cuboid // J. Phys. D: Appl. Phys. 50, 175102 (2017)

41. L. Yue, B. Yan, J. Monks, Z. Wang, N. T. Tung, V. D. Lam, O. V. Minin, and I. V. Minin. A millimetre-wave cuboid solid immersion lens with intensity-enhanced amplitude mask apodization // Journal of Infrared, Millimeter, and Terahertz Waves, 39(6), 546–552 (2018)

42. Minin I.V. and Minin O.V. Brief Review of Acoustical (Sonic) Artificial Lenses // Proc. of the 13th Int. Scientific-technical conf. On actual problems of electronic instrument Engineering (APEIE) -39281, Novosibirsk, Oct.3–6, 2016, v.1, pp.136–137.

43. I.V. Minin, Q. Tang, S. Bhuyan, J. Hu, O.V. Minin. A Method to Improve the Resolution of the Acoustic Microscopy // ArXiv:1712.01638 (2017)

44. I.V. Minin, O. V. Minin and I. S. Tseplyaev. The relationship between resonance scattering and the formation of an acoustojet under the interaction of ultrasound with a dielectric sphere immersed in water // IOP Conf. Series: Journal of Physics: Conf. Series 881, 012025 (2017)

45. I.V. Minin, O. V. Minin, S. E. Shipilov and K. V. Zavyalova. The possibility of total protein concentration determination based on acoustojet phenomenon// IOP Conf. Series: Journal of Physics: Conf. Series 881, 012038 (2017)

46. S. Castineira-Ibanez, D. Tarrazo-Serrano, A. Uris, C. Rubio, O. V. Minin, and I. V. Minin. Cylindrical 3D printed configurable ultrasonic lens for subwavelength focusing enhancement // Sci Rep (under review), (2020)

47. E. Ginzel, R. MacNeil, R. Ginzel, M. Zuber, A. N. Sinclar. Acoustic Properties of the Elastomeric Materials Aqualene and ACE // e-Journal of Nondestructive Testing (NDT), 20(12), 1–12 (2015). https://www.ndt.net/article/ndtnet/2015/9_Ginzel.pdf

48. E. Abbe, Beiträge zur Theorie des Mikroskops und der mikroskopischen Wahrnehmung // Archiv für Mikroskopische Anatomie, 9, pages 413–468 (1873); https://doi.org/10.1007/BF02956173

49. Lord Rayleigh F.R.S. XXXI. Investigations in optics, with special reference to the spectroscope // The London, Edinburgh, and Dublin Philosophical Magazine and Journal of Science, 8(49), 261–274 (1879); https://doi.org/10.1080/14786447908639684

50. Z. Wang, B. Luk'yanchuk, L. Yue, B. Yan, J. Monks, O. V. Minin, I. V. Minin, S. Huang and A. A. Fedyanin. High order Fano resonances and giant magnetic fields in dielectric microspheres // Sci Rep 9, 20293 (2019)

51. L. Yue, B. Yan, J. N. Monks, R. Dhama, C. Jiang, O. V. Minin, I. V. Minin, and Z. Wang. Full three-dimensional Poynting vector analysis of great field-intensity enhancement in a specifically sized spherical-particle // Sci Rep 9, 20224 (2019)

52. L. Yue, Z. Wang, B. Yan, J. Monks, Y. Joya, R. Dhama, O. V. Minin, and I. V. Minin. Superenhancement focusing of Teflon sphere // Annalen der Physik 2000373 (2020)

53. L. Yue, O. V. Minin, Z. Wang, J. Monks, A. Salin, and I. V. Minin, Photonic hook: a new curved light beam // Optics Letters 43(4), 771–774 (2018).

54. I. V.Minin, O. V. Minin, G. Katyba, N. Chernomyrdin, V. Kurlov, K. I. Zaytsev, L. Yue, Z. Wang, and D. N. Christodoulides. Experimental observation of a photonic hook // Appl. Phys. Lett. 114, 031105 (2019)

55. O. V. Minin, I. V. Minin, K. I. Zaytsev, G. Katyba, V. Kurlov, L. Yue, Z. Wang, "Electromagnetic field localization behind a mesoscale dielectric particle with a broken symmetry: a photonic hook phenomenon," Proc. SPIE 11368, Photonics and Plasmonics at the Mesoscale, 1136807 (2 April 2020)

56. C. Rubio, D. Tarrazó-Serrano, O. V. Minin, A. Uris, I. V. Minin. Acoustical hooks: A new subwavelength self-bending beam // Results in Physics 16, 102921 (2020)

57. Cremer L, Heckl M. *Structure-Borne Sound*. 2nd ed. New York: Springer-Verlag; 1988. p. 138.

58. Y. Geynts, A. Zemlyanov, and E. Panina. Features of Photonic Nanojet Formation near Surfaces of Spherical Microparticles Illuminated by a Focused Laser Beam // Atmospheric and Oceanic Optics, 28(2), 139–144 (2015)

59. I. V. Minin, O. V. Minin, C.-Y. Liu, H.-D. Wei, Y. Geints, and A. Karabchevsky. Experimental demonstration of tunable photonic hook by partially illuminated dielectric microcylinder // Optics Letters, 45(17) 4899–4902 (2020)

60. B. Hartmann. Ultrasonic properties of poly(4methyl pentene1) // Journal of Applied Physics 51, 310 (1980)

61. Sinha M., Buckley D.J. *Acoustic Properties of Polymers*. In: Mark J.E. (eds) Physical Properties of Polymers Handbook. Springer, New York, NY (2007).

62. I. Dorofeev, V. Suslyaev, I. V. Minin, O. V. Minin. The role of the tangential electric field component to the terahertz jet and hook formation by dielectric cube and sphere // Opt Eng 60(8), 082004 (2020)

Chapter 4
Plasmonic Hook

Abstract Until recently the surface plasmon polariton (SPP) waves propagating along the curve trajectory were the only Airy-family plasmonic beams. In 2018, a new class of curved surface plasmon wave – the plasmonic hook (PH) – was introduced. The PH is created using the in-plane focusing of the SPP wave through a wavelength-scaled Janus dielectric particle. This is fundamentally simpler than the generation of the known SPP Airy-family beams. The PH propagates along a wavelength-scaled curved trajectory with radius less than the SPP wavelength, which represents the smallest radius of curvature ever recorded for an SPP beam, and can exist despite the strong energy dissipation at metal surface. In this chapter, we discussed some key properties of PH and its experimental verification. Some of applications are also discussed briefly.

4.1 Introduction

It is well known that surface plasmon polaritons (SPPs) have attracted great interest due to their ability to efficiently manipulate electromagnetic wave on a deep sub-wavelength scale. SPP fields are strongly localized at the metal-dielectric interface and thus have a shorter wavelength than the excitation radiation opening up possibilities of subwavelength waveguides, detectors, and enhanced wave-matter interaction [1–4]. SPPs are essentially two-dimensional (2D) waves whose field components decay exponentially with the distance from the surface. One of the key components to excitation and effective control of the SPP is a lens. Different structures based on mirrors, slits, holes, metasurfaces, and dielectric structures mounted on the metal film have become an efficient method for the focusing and manipulating of two-dimensional SPP waves [5–16]. But dielectric plasmonic lenses cannot achieve the manipulation of an SPP energy below its wavelength due to fundamental diffraction limit [13–16].

In [17] it was theoretically for the first time shown that dielectric disks with refractive index of $n = 1.5$ and Mie size parameter $q = 238$ are able to produce localized plasmonic jet (SPPJ) similar to the photonic nanojet (PNJ) effect in optics [18, 19]. Later, this effect was theoretically predicted in a dielectric cuboid particle [20]. The structure comprised a dielectric cuboid of silicon nitride (Si_3N_4) with refractive index $n = 1.97$ and lateral side dimensions equal to $\lambda_0 = 1550$ nm (where λ_0 is a telecommunication wavelength) placed on top of 100-nm-thick gold film. It was shown that the control of the SPPJ can be tailored by changing of the dielectric particle height [17, 20]. It was also demonstrated that the best focusing effect is observed when the height of the dielectric structure is about $0.1\lambda_0$ that allows to observed the SPPJ an intensity enhancement of ~5 at the focus with a resolution of $0.68\lambda_0$ [20].

It must be noted that open-type waveguides, known as Goubau lines, formed by periodically placed lenses have been proposed in [21]. In the cross-section of open-type waveguides, any impenetrable walls do not limit the field in contrast to conventional waveguides. When passing through such a waveguide, the light focuses and defocuses by the periodically located lenses. Such multiple focusing and defocusing help compensating diffraction losses. Based on this idea, the mechanism of SPPJ field localization was proposed to use a chain of dielectric particles [22] in order to extend the SPP propagation length up to ten times [23].

The first experimental observation of the SPPJ effect in the dielectric cuboid mounted on a metal film was reported in [24, 25]. In simulations, as a material for the dielectric cuboid, we used both poly(methyl methacrylate) (PMMA) and AR-P 6200 e-beam resist from Allresist [26] featuring a refractive index of $n = 1.5302$ at $\lambda_0 = 1530$ nm. The relation [15] determined the effective refractive index n_{eff} of the SPP:

$$k_{spp} = k_0 \left(\left(\varepsilon_m \times \varepsilon_d \right) / \left(\varepsilon_m + \varepsilon_d \right) \right) 1/2 = n_{eff} \times k_0,$$

where $k_0 = 2\pi/\lambda_0$ is the wave number in vacuum and ε_d and ε_m are the relative permittivities of dielectric and metal, respectively. From the dispersion relation equation, it is followed that for our conditions, the surface plasmon wavelength is about 2% less than the wavelength in a free space and is $\lambda_{spp} = 0.978\lambda_0 = 1497$ nm.

The simulation results of the SPP field intensity distributions near the dielectric cuboid with three heights ($h = 200$, 250, and 300 nm) and fixed size of 5 μm at $\lambda_0 = 1530$ nm are shown in Fig. 4.1a [24] and correspondent refractive index for two dielectrics (Fig. 4.1b).

When the height of the cuboid changes, the field localization increases linearly, wherein the FWHM value slightly decreases. It means that the effective refractive index n_{eff} of the considered structure depends on the height of the dielectric particle,

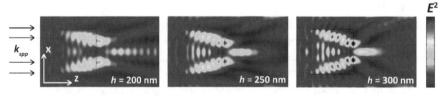

Fig. 4.1a SPP field intensity distributions near the dielectric cuboid placed on metal film with height of $h = 200$, 250, and 300 nm. The surface plasmon wave propagates from left to right [24]

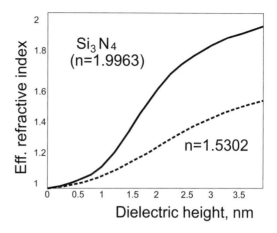

Fig. 4.1b Effective refractive index for two dielectrics, obtained from dispersion relation equation

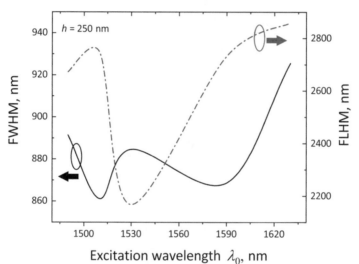

Fig. 4.2 Simulation results of the FWHM (solid curve) and FLHM (dash-dotted curve) of the SPPJ propagation vs the excitation wavelength [24]

as expected [13–17, 20, 24, 25]. The length of SPPJ defined as a full length at half maximum (FLHM) along SPP propagation and FWHM at different excitation wavelengths are shown in Fig. 4.2 [24]. One can see that the FWHM is about $0.58\lambda_0$ and the FLHM is about $1.41\lambda_0$ at $\lambda_0 = 1530$ nm.

In Fig. 4.3, the experimental image of the SPPJ propagation derived using s-SNOM is shown [24]. A low divergence of plasmonic jet and high-intensity sub-wavelength spots at the communication wavelength of $\lambda_0 = 1530$ nm by using a rather simple dielectric microstructure are clearly visible. Thus in [24, 25] for the first time, we have experimentally demonstrated that a simple dielectric wavelength-

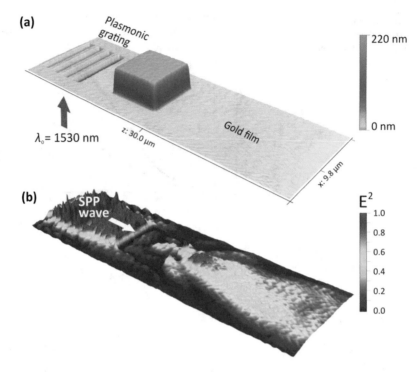

Fig. 4.3 Experimental image of the SPPJ propagation derived using s-SNOM. The optical excitation of $\lambda_0 = 1530$ nm is incident on the backside of the plasmonic grating resulting in the excitation of SPP waves in the orthogonal direction. Reprinted from [25] with permission © The Optical Society

scaled particle [27–33] combined to a metal film can act as plasmonic focusing structure (Fig. 4.4) and thus can focus surface plasmons in a subwavelength scale.

The effect of plasmonic analogue of optical photonic jet is believed to show exciting potential applications in near-field and integrated optics to control and guide SPPs at subwavelength scale and a relatively cheap and simple way for the manipulation of the SPP propagation even below the diffraction limit.

Such simple and novel platform can provide new pathways for plasmonics, strong-field nano-optics [34], and high-resolution imaging and serves as the basis for the formation of plasmonic hooks.

4.2 Plasmonic Hook

Salandrino and Christodoulides have suggested Airy beams theoretically in plasmonics in 2010 [35]. A unique feature of the Airy beam is that it maintains its shape even in a 2D case, which consists of a transverse and the propagation coordinate. The discovery of self-bending Airy-like beams for SPP has driven new opportuni-

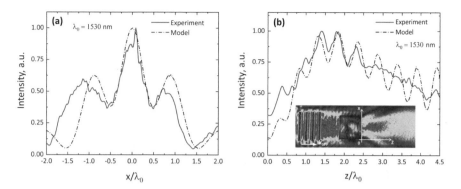

Fig. 4.4 SPP field intensity distributions near the shadow side of the dielectric particle on the gold film (**a**) in the focal spot and (**b**) along SPP jet propagation. The solid curves correspond to the experiment, while the dashed-dotted curves refer to the modeling. The inset in (**b**) depicts the SNOM image of intensity distributions. Reprinted from [25] with permission © The Optical Society

ties for different area of plasmonics, imaging, sensing. Among them are Bessel [36–40], as well as Weber and Mathieu SPP beams [41, 42]. Until recently, the Airy-like SPP family beams, which were realized in a low-dimensional system, have been the only plasmonic waves that are self-healing and quasi-diffraction-free. Thus they are unaffected by any structural imperfections and in particular surface roughness [43], which is important in plasmonics. It could be noted that the Airy-family SPP beams in contrast to classical 2D non-diffracting beams [44] are used to be non-diffracting wave packets in one-dimensional (1D) planar systems [45, 46].

In order to realize the unique curved hook beams as SPP waves, several fundamental challenges, owing to the plasmonic nature of the waves, should be addressed. For example, due to the limited propagation length of SPP, the resulting SPP jets and hooks should be formed directly in the near-field (and therefore their phase cannot be simply defined at a specific 1D plane), before they decay. Moreover, Airy-like SPP require complex techniques to compensate a wave vector mismatch between SPP and a free-space wave [38]. To overcome these fundamental challenges, we introduce a new class of SPP curved beams – a plasmonic hook (PH), which are designed specifically for the near-field and that could be realized when the in-plane SPP wave is passing through the Janus dielectric mesoscale particle [47]. This is fundamentally simpler than the generation of the classical SPP Airy-family beams.

In simulations [47] a light source with wavelength $\lambda_0 = 800$ nm was used to SPP excitation. Refractive index of dielectric structure was 1.35. The height of the dielectric and gold film was $h = 600$ and 100 nm, respectively (see Fig. 4.5).

It has been shown that the self-bending structure of the PH arises from constructive interference between the incident, diffracted, and scattered fields near the dielectric structure due to the near-field interactions at the shadow surface of the dielectric particle (see Fig. 4.6) and can exist despite the strong energy dissipation

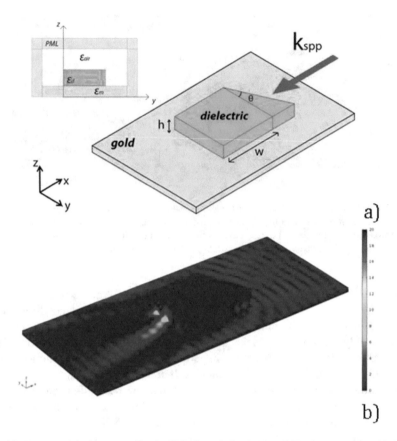

Fig. 4.5 Janus particle for generation the SPP photonic hook (**a**) and 3D view together with field intensity distributions near the dielectric cube 2.5λspp×2.5λspp with prism angle θ = 23.6°. (**b**). Reprinted from [47] with permission by John Wiley and Sons

arising from large Ohmic losses in noble metals. It was shown that the PH propagates along a wavelength-scaled curved trajectory with a radius less than the SPP wavelength compared to the classical Airy-family SPP beams [43, 45, 46], representing the smallest radius of curvature ever recorded for an SPP beam before. Importantly, the specific curved shape of the hook may appear even in free space and requires no any external potential [48, 49] or waveguiding structures [14, 50–54].

From the simulations [47], it was also followed that the minimum SPP beamwidth (in terms of FWHM, full width at half maximum [55]) corresponds to the bending point of the PH and for PH is smaller than for plasmonic jet caused by a symmetric cubic particle. And, thus, PH is able to break the simplified (0.5λ criterion [56, 57]) diffraction limit.

Moreover, it has been shown that the PH exiting at different wavelengths will propagate along trajectories with different orientation and curvatures, and thus structure of the PH may be changed by the incident light by means of fixed coupling

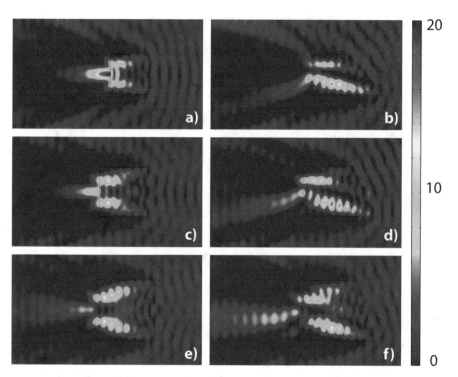

Fig. 4.6 Simulations of SPP field intensity distributions for different regular and Janus dielectric particles: (**a**) 2.0λspp $\times$ 2.0λspp, $\theta = 0$, (**b**) 2.0λspp $\times$ 2.0λspp, $\theta = 44.4$, (**c**) 2.5λspp $\times$ 2.5λspp, $\theta = 0$, (**d**) 2.5λspp $\times$ 2.5λspp, $\theta = 23.6$, (**e**) 3.0λspp $\times$ 3.0λspp, $\theta = 0$, and (**f**) 3.0λspp $\times$ 3.0λspp, $\theta = 15.5$ (θ in degree). Reprinted from [47] with permission by John Wiley and Sons

Janus particle (Fig. 4.7) and might be adopted to dynamically control 2D SPP curved beams [47].

In 2019, I.V. Minin and O.V. Minin et al. experimentally verified the photonic hook phenomenon in free space [51, 52, 58]. As in [25] to verify SPP PH effect [59], we use dielectric Janus particle, which was placed on a 100-nm-thick gold film and made from AR-P 6200 e-beam resist (Allresist GmbH) [26]. In the simulation, the SPP wave being excited at the telecom wavelength $\lambda_0 = 1530$ nm. The Janus particle has the lateral dimensions $l_x = l_z = h = 5085$ nm, and the prism angle equals to $\theta = 27°$. We use a Drude-Lorentz dispersion model with relative permittivity of gold $\varepsilon_m = -114.47 + 8.51i$ at $\lambda_0 = 1530$ nm [60, 61]. As expected, simulations have shown that formation of the hook is severely dependent on the cube height, and according to the simulation results, the maximum curvature of the SPP PH was observed at $h = 250$ nm for the wavelength of 1530 nm.

The electron-beam lithography was used for Janus particle fabrication [59], which was placed at 5 μm from the last slit in the grating. The scanning near-field optical microscopy measurements were performed using the neaSNOM from Neaspec GmbH (s-SNOM) [62].

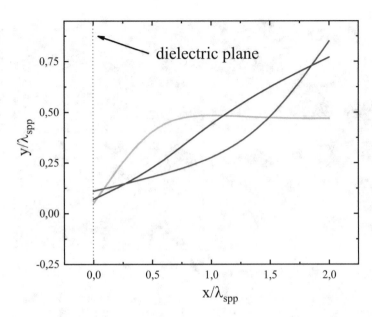

Fig. 4.7 Profile of SPP PH for dielectric height h_{diel} = 1000 nm and illuminated wavelength λ_0 = 700 (green), 800 (blue), and 900 (red) nm, respectively. Reprinted from [47] with permission by John Wiley and Sons

The experimental results are shown in Fig. 4.8 [59]. As follows from Fig. 4.8a, the plasmonic localized beam begins bending starting from the inflection point. This demonstrates the PH effect. Figure 4.8b and c illustrate cross-sections of the SPP field intensity distributions near the shadow surface of the Janus particle in a focal spot and along the PH propagation axis, respectively.

From the experimental data, it is followed that the curvature of the SPP PH is $\beta = 16°$ at λ_0 = 1530 nm. Importantly, in the SPP PH there is an inflexion where the curved beam changes its propagating direction. This property is not possessed by the Airy-like beam [63]. One can see in Fig. 4.8a that the beam starts bending at the distance about the wavelength of optical excitation. The results briefly described above for the first time experimentally show the smallest radius of curvature ever recorded for SPP waves compared to that for known before the Airy-family plasmonic beams.

Conclusion The effects of the SPP hook and SPP jet, which briefly discussed in this Chapter, open up a new degree of freedom for controlling and manipulating surface plasmons polaritons by simple dielectric Janus particles, placed on the surface on the metal film. These both effects (as the SPP jet as the SPP hook) are based on using a variation of the plasmonic wave-vector by the variable dielectric

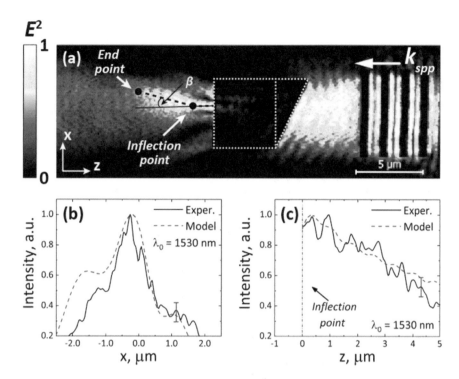

Fig. 4.8 SNOM images illustrating formation of the PH and high-intensity spots (**a**); cross-sections of the SPP field intensity distributions near the shadow side of the Janus particle in a focal spot (**b**) and along the localized beam propagation axis (**c**). The SPP wave is incident on the Janus particle from right to left (**a**). The color bar demonstrates the intensity profile of the SPP PH propagation [59]

thickness. Both effects as SPP jet as PH do exhibit subwavelength field localization, making them possible to use in highly integrated optical circuits. Moreover, experimentally observed PH demonstrated the smallest radius of curvature ever recorded for surface plasmon waves providing many useful applications. As example we believe that the proposed approach can be potentially applied to other types of surface waves such as Bloch surface waves [64, 65], Dyakonov surface waves [66, 67], and so on [68]. Apart from being interesting in their own right, PHs may also hold promise for new, exciting applications in the general area of optoplasmonics [69]; plasmonics, including SPP optical tweezers [70–72]; for on-chip communications [73]; and for nano-scale light switching and guiding in a photonic integrated chip. Formation of SPP PH allows to do the local control of SPP propagation and, in consequence, allows the possibility to obtain SPP localization in a controlled curvilinear form, etc. This field is in its infancy, with various methods and structures being currently investigated.

References

1. D. Bohm and E. P. Gross. Theory of Plasma Oscillations. A. Origin of Medium-Like Behavior // Phys. Rev. 75, 1851 (1949)
2. E. Ozbay. Plasmonics: Merging Photonics and Electronics at Nanoscale Dimensions // *Science,* 311(5758), 189–193 (2006)
3. Z. Liu, J. Steele, W. Srituravanich, Y. Pikus, C. Sun, X. Zhang, Focusing Surface Plasmons with a Plasmonic Lens // Nano Lett. 5 (9), 1726–1729 (2005).
4. J. A. Polo, Jr., T. G. Mackay, and A. Lakhtakia, *Electromagnetic Surface Waves: A Modern Perspective* (Elsevier, Waltham, MA, 2013).
5. N. C. Lindquist, P. Nagpal, A. Lesuffleur, D. J. Norris, S.-H. Oh. Three-Dimensional Plasmonic Nanofocusing // Nano Lett. 10, 1369–1373 (2010).
6. I. V. Minin, O.V. Minin, 3D diffractive focusing THz of in-plane surface plasmon polariton waves // J. of Electromagnetic Analysis and Applications, 2, 116–119 (2010)
7. G. Wu, J. J. Chen, R. Zhang, J. H. Xiao, Q. H. Gong, Highly Efficient Nanofocusing in a Single Step-like Microslit // Opt. Lett. 38 (19), 3776–3779 (2013).
8. M. Takeda, S. Okuda, T. Inoue, K. Aizawa, Focusing Characteristics of a Spiral Plasmonic Lens // Jpn. J. Appl. Phys. 52, 09LG03 (2013).
9. R. G. Mote, O. V. Minin, I.V. Minin, Focusing behavior of 2-dimensional plasmonic conical zone plate // Opt. Quant. Electron. 49, 271 (2017).
10. P. Melentiev, A. Kuzin, D. Negrov, V. Balykin, Diffraction-Limited Focusing of Plasmonic Wave by a Parabolic Mirror // Plasmonics, 13 (6), 2361–2367 (2018).
11. X. Yin, T. Steinle, L. Huang, T. Taubner, M. Wuttig, T. Zentgraf, H. Giessen, Beam switching and bifocal zoom lensing using active plasmonic metasurfaces // Light: Science & Applications, 6, e17016 (2017).
12. V. Smolyaninova, I. Smolyaninov, A. Kildishev, V. Shalaev, Maxwell fish-eye and Eaton lenses emulated by microdroplets // Opt. Lett. 35, 3396–3398 (2010).
13. A. Hohenau, J. Krenn, A. Stepanov, A. Drezet, H. Ditlbacher, B. Steinberger, A. Leitner, F. Aussenegg, Dielectric optical elements for surface plasmons // Opt. Lett., 30 (8), 893–895 (2005).
14. I. V. Minin, O. V. Minin, *Diffractive Optics and Nanophotonics: Resolution Below the Diffraction Limit*, Springer, Cham (2015).
15. W. Shi, T. Chen, H. Jing, R. Peng, M. Wang, Dielectric lens guides in-plane propagation of surface plasmon polaritons // Opt. Express, 25 (5), 5772–5780 (2017).
16. C. Gartcia-Ortiz, R. Cortes, J. Gomez-Correa, E. Pisano, J. Fiutowski, R. Garcia-Ortiz, V. Ruitz-Cortes, H. Rubahn, V. Coello. Plasmonic metasurface Luneburg lens // Photonics Research, 7(10), 1112 (2019).
17. D. Ju, H. Pei, Y. Jiang, X. Sun, Controllable and enhanced nanojet effects excited by surface plasmon polariton // Appl. Phys. Lett. 102, 171109 (2013).
18. A. Heifetz, S. Kong, A. Sahakian, A. Taflove, V. Backman, Photonic Nanojets // Journal of Computational and Theoretical Nanoscience, 6/9, 1979–1992 (2009).
19. B. Luk'yanchuk, R. Paniagua-Domínguez, I. V. Minin, O.V. Minin, Z. Wang, Refractive index less than two: Photonic nanojets yesterday, today and tomorrow // Opt. Mater. Express, 7, 1820–1847 (2017).
20. V. Pacheco-Pena, I. V. Minin, O.V. Minin, M. Beruete, Comprehensive analysis of photonic nanojets in 3D dielectric cuboids excited by surface plasmons // Ann. Phys. 528, 1–9 (2016).
21. G. Goubau, F. Schwering, On the guided propagation of electromagnetic wave beams // IRE Trans. Antennas Propag. 9, 248-256 (1961).
22. I. V. Minin, O.V. Minin, V. Pacheco-Peña, M. Beruete. All-dielectric periodic terajet waveguide using an array of coupled cuboids // Appl. Phys. Lett. 106, 254102 (2015).

23. V. Pacheco-Pena, I. V. Minin, O.V. Minin, M. Beruete, Increasing Surface Plasmons Propagation via Photonic Nanojets with Periodically Spaced 3D Dielectric Cuboids // Photonics, 3, 1–7 (2016).

24. I. V. Minin, O. V. Minin, D. Ponomarev, I. Glinskiy, D. Yakubovsky, V. Volcov. First experimental observation of plasmonic photonic jet based on dielectric cube // arXiv:1912.13373 (2019)

25. I. V. Minin, O. V. Minin, I. Glinskiy, R. Khabibullin, R. Malureanu, A. Lavrinenko, D. Yakubovsky, A. Arsenin, V. Volkov, and D. Ponomarev. Plasmonic nanojet: an experimental demonstration // Optics Letters 45(12), 3244 (2020)

26. Allresist GmbH Datasheet. "Positive E-Beam Resists AR-P 6200 (CSAR 62)," https://www.nanophys.kth.se/nanolab/resists/allresist/produktinfos_ar-p6200_englisch.pdf

27. V. Pacheco-Peña, M. Beruete, I. V. Minin, O.V. Minin, Terajets produced by dielectric cuboids // Appl. Phys. Lett. *105*, 084102 (2014).

28. O. V. Minin, I.V. Minin, Terahertz artificial dielectric cuboid lens on substrate for super-resolution images // Opt. Quantum Electron. *49*, 326–329 (2017).

29. M. Khodzinsky, A. Vosianova, V. Gill, A. Chernyadiev, A. Grebenchukov, I. V. Minin, and O. V. Minin. "Formation of terahertz beams produced by artificial dielectric periodical structures", Proc. SPIE 9918, Metamaterials, Metadevices, and Metasystems 2016, 99182X (16 September 2016)

30. I.V. Minin, C.-Y. Liu, Y. E Geints, O. V. Minin. Recent advantages in Integrated Photonic Jet-Based Photonics // Photonics 7(2), 41 (2020)

31. H. Pham, S. Hisatake, I. V. Minin, O. V. Minin, and T. Nagatsuma. Three-Dimensional Direct Observation of Gouy Phase Shift in a Terajet Produced by a Dielectric Cuboid // APL, 108, 191102 (2016)

32. H. Pham, S. Hisatake, O.V. Minin, T. Nagatsuma and I.V. Minin, Asymmetric Phase Anomaly of Terajet Generated from Dielectric Cube under Oblique Illumination // Appl. Phys. Lett. 110(20), 201105 (2017)

33. I. V. Minin, C.-Y. Liu, Y.-C. Yang, K. Staliunas and O. V. Minin. Experimental observation of flat focusing mirror based on photonic jet effect // Sci Rep 10, 8459 (2020)

34. P. Dombi, Z. Pápa, J. Vogelsang, S. Yalunin, M. Sivis, G. Herink, S. Schafer, P. Groß, C. Ropers and C. Lienau. Strong-field nano-optics // Rev. Mod. Phys., 92(2), 025003 (2020)

35. A. Salandrino and D.N. Christodoulides. Airy plasmon: a nondiffracting surface wave // Opt. Lett. 35, 2082 (2010).

36. H. Kano, D. Nomura, H. Shibuya, Excitation of surface-plasmon polaritons by use of a zeroth-order Bessel beam // Appl. Optics 43, 2409 (2004).

37. C. E. Garcia-Ortiz, V. Coello, S.T. Bozhevolnyi, Generation of diffraction-free plasmonic beams with one-dimensional Bessel profiles // Opt. Lett. 38, 905 (2013).

38. A. Minovich, A. E. Klein, N. Janunts, T. Pertsch, D. N. Neshev, and Y. S. Kivshar. Generation and Near-Field Imaging of Airy Surface Plasmons // Phys. Rev. Lett. 107, 116802 (2011).

39. L. Li, T. Li, S.M. Wang, C. Zhang, and S.N. Zhu. Plasmonic Airy Beam Generated by In-Plane Diffraction // Phys. Rev. Lett. 107, 126804 (2011).

40. X. Song, L. Huang, L. Sun, X. Zhang, R. Zhao, X. Li, J. Wang, B. Bai, and Y. Wang. Near-field plasmonic beam engineering with complex amplitude modulation based on metasurface // Applied Physics Letters 112(7):073104 (2018)

41. A. Libster-Hershko, I. Epstein, A. Arie, Rapidly accelerating Mathieu and Weber surface plasmon beams // Phys. Rev. Lett. 113, 123902 (2014).

42. I. Epstein, H. Suchowski, D. Wesiman, R. Remez, and A. Arie. Observation of linear plasmonic breathers and adiabatic elimination in a plasmonic multi-level coupled system // Opt. Express, 26(2), 1433 (2018)

43. N. Efremidis, Z. Chen, M. Segev, D. N. Christodoulides, Airy beams and accelerating waves: An overview of recent advances // Optica, *6*, 686–701 (2019).

44. P. Zhang, S. Wang, Y. Liu, X. Yin, C. Lu, Z. Chen, X. Zhang, Plasmonic Airy beams with dynamically controlled trajectories // Optics Letters, 36(16), 3191–3193 (2011)
45. J. Durnin, J.J. Miceli Jr., J. H. Eberly, Diffraction-free beams // Phys. Rev. Lett. 58, 1499 (1987).
46. G. Siviloglou, J. Broky, A. Dogariu, D. N. Christodoulides, Observation of accelerating Airy beams // Phys. Rev. Lett. 99, 213901 (2007).
47. I.V. Minin, O.V. Minin, D.S. Ponomarev, I.A. Glinskiy, Photonic Hook Plasmons: A New Curved Surface Wave // Ann. Physik. 1800359 (2018)
48. W. Liu, D.N. Neshev, I.V. Shadrivov, A.E. Miroshnichenko, Y.S. Kivshar, Plasmonic Airy beam manipulation in linear optical potentials // Opt. Lett. 36, 1164 (2011).
49. N. K. Efremidis, Airy trajectory engineering in dynamic linear index potentials // Opt. Lett. 36, 3006– 3008 (2011).
50. L. Yue, O. V. Minin, Z. Wang, J. Monks, A. Salin, and I. V. Minin, Photonic hook: a new curved light beam // Optics Letters 43(4), 771–774 (2018).
51. I. V. Minin, O. V. Minin, G. Katyba, N. Chernomyrdin, V. Kurlov, K. I. Zaytsev, L. Yue, Z. Wang, and D. N. Christodoulides. Experimental observation of a photonic hook // Appl. Phys. Lett. 114, 031105 (2019)
52. O. V. Minin, I. V. Minin, K. I. Zaytsev, G. Katyba, V. Kurlov, L. Yue, Z. Wang, "Electromagnetic field localization behind a mesoscale dielectric particle with a broken symmetry: a photonic hook phenomenon," Proc. SPIE 11368, Photonics and Plasmonics at the Mesoscale, 1136807 (2 April 2020)
53. I. V. Minin, O. V. Minin, L. Yue, Z. Wang, V. Volcov, and D. N. Christodoulides. Photonic hook – a new type of subwavelength self-bending structured light beams: a tutorial review // ArXiv: 1910.09543 (2019)
54. A.S. Ang, A. Karabchevsky, I.V. Minin, O.V. Minin, S.V. Sukhov, and A.S. Shalin. 'Photonic Hook' based optomechanical nanoparticle manipulator // Scientific Reports 8, 2029 (2018)
55. W. V. Houston, A compound interferometer for fine structure work // Phys Rev, 29, 0478–0484 (1927)
56. E. Abbe, Beiträge zur Theorie des Mikroskops und der mikroskopischen Wahrnehmung // Archiv für Mikroskopische Anatomie, 9, pages 413–468 (1873); https://doi.org/10.1007/BF02956173
57. Lord Rayleigh F.R.S. XXXI. Investigations in optics, with special reference to the spectroscope // The London, Edinburgh, and Dublin Philosophical Magazine and Journal of Science, 8(49), 261–274 (1879); https://doi.org/10.1080/14786447908639684
58. K. Dholakia, G. Bruce, Optical hooks, // Nature Photonics 13(4), 229–230 (2019).
59. I. V. Minin, O. V. Minin, I. Glinskiy, R. Khabibullin, R. Malureanu, D. Yakubovsky, V. Volkov, D. Ponomarev. Experimental verification of a plasmonic hook in a dielectric Janus particle // arXiv:2004.10749 (2020)
60. A. Vial, T. Laroche, M. Dridi, L. Le Cunff. A new model of dispersion for metals leading to a more accurate modeling of plasmonic structures using the FDTD method. // Appl Phys A 103(3), 849–853 (2011).
61. H. S. Sehmi, W. Langbein, and E. A. Muljarov. Optimizing the Drude-Lorentz model for material permittivity: Method, program, and examples for gold, silver, and copper // Phys. Rev. B 95, 115444 (2017)
62. F.J. Alfaro-Mozaz, P. Alonso-González, S. Vélez, I. Dolado, M. Autore, S. Mastel, F. Casanova, L.E. Hueso, P. Li, A.Y. Nikitin and R. Hillenbrand, Nanoimaging of resonating hyperbolic polaritons in linear boron nitride antennas // Nature Communications, 8, 15624 (2017)
63. I. V. Minin and O. V. Minin. Recent Trends in Optical Manipulation Inspired by Mesoscale Photonics and Diffraction Optics // J of Biomedical Photonics & Eng 6(2), 020301 (2020)
64. A.P.Vinogradov, A.V.Dorofeenko, A.M.Merzlikin, and A.A.Lisyansky, Surface states in photonic crystals // Phys.Usp. 53, 243 (2010).
65. L. Yu, E. Barakat, T. Sfez, L. Hvozdara, J. DiFrancesco, and H.P. Herzig, Manipulating Bloch surface waves in 2D: a platform concept-based flat lens // Light 3, e124 (2014).

66. M.I. Dyakonov, New type of electromagnetic wave propagating at an interface // Sov.Phys.J. 67, 714 (1988).
67. O.Takayama, D.Artigas, and L.Torner, Lossless directional guiding of light in dielectric nanosheets using Dyakonov surface waves // Nat.Nanotechnol. 9, 419 (2014).
68. J.A. Polo Jr., and A. Lakhtakia. Surface electromagnetic waves: A review // Laser Photonics Rev. 5, 2, 234–246 (2011)
69. Y. Hong and B. M Reinhard. Optoplasmonics: basic principles and applications // J. Opt. 21, 113001 (2019)
70. M. Righini, G. Volpe, C. Girard, D. Petrov, and R. Quidant. Surface Plasmon Optical Tweezers: Tunable Optical Manipulation in the Femtonewton Range // PRL 100, 186804 (2008)
71. M. L. Juan, M. Righini and R. Quidant. Plasmon nano-optical tweezers // Nature Photonics, 5, 349 (2011)
72. J.-S. Huang and Y.-T. Yang. Origin and Future of Plasmonic Optical Tweezers // Nanomaterials 5, 1048–1065 (2015)
73. T. W. Ebbesen, C. Genet, and S. I. Bozhevolnyi, Surface plasmon circuitry // Phys. Today 61, 44–50 (2008).

Chapter 5
Photonic Hook Effect Applications

Abstract Specialized electromagnetic fields can be used for nanoparticle manipulation along a specific path, allowing enhanced transport and control over the particle's motion. In this chapter we show that a gold nanoparticle immersed in the Janus dielectric particle's transmitted field moves in a subwavelength curved trajectory. This result could be used for moving nanoparticles around obstacles. However, curved forces such as photonic hooks are extremely weak in low-contrast media. Here, we discuss the amplification of optical forces generated by a photonic hook via pulsed illumination mediated by temperature effects. We show that the optical force generated by the photonic hook subjected to illumination by an incident Gaussian pulse is significantly larger than the optical force generated by the photonic hook subjected to a continuous wave. Several other promising applications of the hook phenomenon are also being discussed: optical switch, optical vacuum cleaner, optical scalpel, photonic hook in reflection mode, acoustical trapping, and surface hydroelastic waves.

5.1 Introduction

Electromagnetic fields exert forces on particles, and by using specialized electromagnetic fields, it is possible to control a nanoparticle's motion over a particular trajectory. Below we briefly investigate the electromagnetic forces produced by a "photonic hook" created by a dielectric cuboid with broken symmetry, formed by appending a triangular wedge to one of face of a cube. This simplicity in our construction gives our system an advantage over conventional methods of generating curved beams, such as Airy or Bessel beams, and the size of our system means that it can also be integrated into lab-on-chip platforms.

© The Author(s), under exclusive license to Springer Nature Switzerland AG 2021 69
O. V. Minin, I. V. Minin, *The Photonic Hook*, SpringerBriefs in Physics,
https://doi.org/10.1007/978-3-030-66945-4_5

5.2   Near-Field Nanoparticle Manipulations

It is well known that dielectric mesoscale particles with refractive index less than 2 in transmission mode allow the creation of a so-called photonic nanojet (PNJ), a highly localized, subwavelength, low-divergence beam [1, 2]. This system could be considered as a perspective micro system for optical manipulation of biological objects [3, 4], and in optical tweezers, special optical powers induced by the structured field of the classical PNJ [5–10] can be used. It is noteworthy that the above-mentioned studies on PNJ-induced optical forces focused on trapping particles along the axis of symmetry.

In the pioneer work [11], we have offered to use a photonic hook (PH) effect for moving nanoparticles around obstacles. A PH is a structured field formed by a Janus mesoscale particle with broken in shape symmetry [12] – as a method of generating optical forces for moving particles in a curved trajectory was later described by simulations in details in [13, 14]. Forces generated by Airy-like beams have been studied earlier [see, e.g., 15, 16], but the generation of such types of beams is usually achieved either using complex, expensive, and demanding methods [17] or using a special telescopic system [18]. Moreover, such beams formed at distances significantly exceeding the wavelength of the incident field [16]. The presence of cumbersome optical elements makes conventional Airy-family beam implementations unsuitable for a variety of laboratory-on-chip applications and platforms. On the other hand, a photonic hook can be created using a compact optical element with a wavelength-scaled dimensions and broken symmetry [11–14, 19].

The Janus particle geometry is shown in Fig. 5.1 [13]. The particle was made from fused silica and illuminated by a plane wave with a wavelength of 625 nm, propagating along the x-axis, and polarized along the y-direction. The detailed description of the numerical simulation can be found in [13].

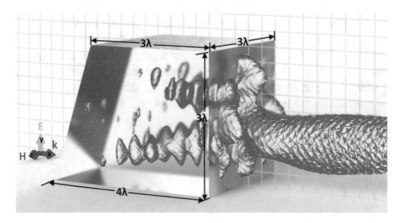

Fig. 5.1 Janus particle geometry. Reprinted from [13] under the Creative Commons Attribution License

For a particle to be moved by a radiation pressure along the power flow in a PH (PNJ), it is necessary to use particles consisting of a material with sufficient imaginary part of dielectric permittivity (e.g., a metallic particle). It follows from the simple formula for dipolar particle [20]:

$$\mathbf{F}_{EM} = \frac{\alpha'}{4}\nabla|\mathbf{E}|^2 + \frac{\alpha''}{2}|\mathbf{E}|^2\nabla\varphi,$$

where φ is the field phase and $\alpha = \alpha' + i\alpha''$ is the complex particle's polarizability [21] and $\mathbf{E}$ – electric field – acting on a particle. The first summand corresponds to the so-called gradient optical force, while the second one is the "radiation pressure," pushing particle in the direction of phase gradient (which coincides with Poynting vector in our case).

In [13] we have considered the interesting application of a PH for guiding particles around obstacles made from gold and glass slabs placed above the hook's path. The obstacles were placed at the distance of maximum curvature of the hook from the shadow surface of Janus particle (about of half of wavelength). It was discovered an interesting effect: the trajectory of the particle in the region x >2500 nm lays above the y axis for the glass obstacle of $\lambda/2$ thickness, and the trajectory nearly coincides with the y axis in the case of glass obstacle of $\lambda/4$ thickness, shown in Fig. 5.2c and d, respectively. The phase difference produced by the field as it goes through the glass obstacle creates the new PNJ, shown in Fig. 5.2a. From Fig. 5.2, it is followed that in the case of the glass obstacle, the particle will move around an obstacle with $\lambda/4$ thickness, but for a $\lambda/2$-thick glass obstacle, the particle will be trapped. The field between the cuboid and the obstacle, shown in Fig. 5.2b, has an appearance similar to the interference pattern of a standing wave, which have been previously studied [9, 10, 14].

Thus we demonstrate that a particle could be stably trapped near the glass obstacle [13] or even go around it [11], which allows new applications in optical tweezing as well as in making nano-manipulation more flexible. A gold nanoparticle immersed in the Janus particle's transmitted field moves in a curved trajectory without the complicated employment of Airy beams. We show that despite the obstacles, a nanoparticle can move around glass obstacles of a specific thickness. For larger glass slabs, the particle will be trapped stably near it.

The possibility to bend the nanoparticle trajectory (without using Airy beams) around transparent obstacles at the subwavelength level without changing an optomechanical system opens a new way of opportunities for optical, biological, and mechanical research as well as applications, including on-chip microfluidics for subwavelength manipulations. We explore one possible application of this phenomenon as an optical-based particle sorter; this would require embedding the cuboid atop a substrate [22]. However, in this setup, scattering from the substrate generates a set of undesired trapping points along the original trajectory. To resolve this, we added an approximately index-matched fluidic background. Our simulations show that, with enough modifications to the Janus particle's material and dimensional parameters, the photonic hook system can be reconfigured to work while embedded on a substrate.

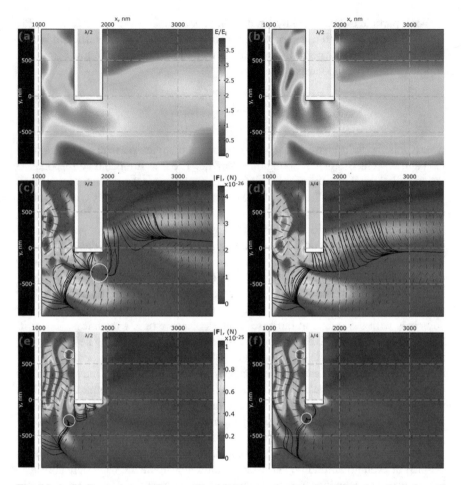

Fig. 5.2 (**a, b**) Comparison of the normalized field strength when the (**a**) λ/2 glass thickness and (**b**) λ/2 gold obstacles are introduced. (**c–f**) Forces produced by the cuboid system with an additional (**c, d**) glass obstacle with (**c**) width λ/2 and (**d**) width λ/4 and a (**e, f**) gold obstacle with (**e**) width λ/2 and (**f**) width λ/4. The color plot represents the force magnitude, the streamlines and arrows represent possible trajectories and the force direction, respectively. White circles in (**c**), (**e**), and (**f**) show the locations of trapping. Figures (**a**) and (**b**); (**c**) and (**d**); and (**e**) and (**f**) use the same color scales (trapping position is indicated with white circle). Reprinted from [13] under the Creative Commons Attribution License

Additional, because an absolute trapping forces depend on the details of the type of illuminated wave (CW or pulsed [23]) and the beam structure, in [24] we study the PH effect generated by a Gaussian pulse and its influence on the temporal dynamics of electronic and lattice temperature.

Simulations (Fig. 5.3) show that the structure of the optical forces for pulsed radiation in the region of the PH resembles a typical picture of standing waves [10], located along a photonic hook field, in contrast to CW illumination. It has been

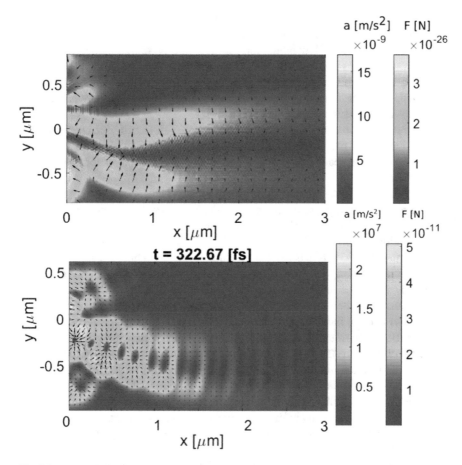

Fig. 5.3 Simulations of acceleration, *a*, of gold nanoparticle and forces acting on a spherical gold nanoparticle with a radius of 30 nm due to (top) CW illumination and (bottom) pulsed illumination with a wavelength of 500 nm. The arrows indicate the direction of the optical force. Reprinted from [24], published by The Royal Society of Chemistry (RSC) on behalf of the Centre National de la Recherche Scientifique (CNRS) and the RSC

shown that heating effects can also change the dielectric permittivity of the particle, thus enabling significantly larger optical forces. Specifically, the acceleration and the force for CW illumination are 15 orders of magnitude less as than for pulsed illumination [24]. It was also discovered that the spectral location of the maximum value of the Purcell factor is shifted due to the heating effect of the "hot" gold nanoparticle as compared to the "cold" one. Femtosecond optical tweezers may be also used to significantly enhance two-photon fluorescence compared to CW illumi-nation [23].

Thus, the velocity and the magnitude of the optical force acting on the gold nanoparticle subject to the pulsed illumination are greater than the force and velocity of the same nanoparticle under CW illumination and proposed a novel tracking concept for particle movement in low-contrast media. In the development of these ideas, the photonic hook applied on a titanium carbide nanoparticle (MXene) for the particle's optomechanical manipulation for the first time was studied in [25]. It has been shown that in comparison with nanoparticles made of gold, experiencing localized surface plasmon resonance in the visible, this resonance of MXene nanoparticles can be tuned to the near-infrared region. These finding are particularly important for biomedical applications.

It is interesting to note that such bendable wavelength-scaled light beams instead of traditional Gaussian beam for line-of-sight light communications may be used for in-plane or free-space data-carrying bendable light communications along arbitrary trajectories.

The "photonic hook" effect may be used for nano-scale light switching and guiding, for example, in a photonic integrated chip. For example, a size-fixed dielectric Janus particle should function like an optical switch for light-route selection depending on incident light wavelength [26]. It is proved that the incident light with a wavelength that is 33% of particle length (short wavelength light in Fig. 5.4) significantly deviates its original straight propagation route and creates a photonic hook with a 35° deflection angle after passing the particle. Otherwise, the same particle cannot bend the light with a wavelength as long as its length (long wavelength light in Fig. 5.4) and the incident light maintain its straight propagation route as a photonic jet. Based on this phenomenon, a dielectric Janus particle can simultaneously guide two lights varying in wavelength to the different light route in the form of photonic jet or photonic hook for frequency (wavelength) switching, as shown in Fig. 5.4 [26]. A similar effect also observed for the masked symmetric particles discussed in Chap. 2. This approach paves the way to hook-based multiplexing circuits enabling simultaneous information transport and processing in separated frequency channels and may be used in various other wave-based applications such as in fiber-optic networks.

The concept of an "optical vacuum cleaner" for optomechanical manipulation of nanoparticles based on nanostructured mesoscale particles of one wavelength dimensions with refractive index contrast near 2 was offered in [27–29] (See Fig. 5.5)

Numerical simulation shows that light can be confined inside the nanohole of the proposed nanostructured dielectric particle, when the hole size is as small as $\lambda/(40...100)$. Importantly the spatial dimensions of the field localization area are

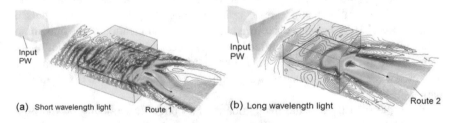

(a) Short wavelength light   Route 1   (b) Long wavelength light   Route 2

Fig. 5.4 Diagram of a particle optical-switch (**a**) short wavelength light – 33% of particle length (**b**) long wavelength light – 100% of particle length. Reprinted from [26]

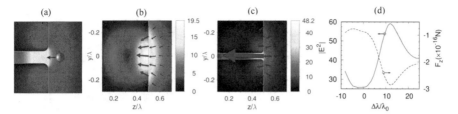

Fig. 5.5 (**a**) Schematic diagram for the "optical vacuum cleaner," where a nanoparticle is pulled by the optical force and moves toward the nanohole structured particle. (**b, c**) Electric field intensity and optical force distributions for (**b**) solid cuboid without a hole, (**c**) cuboid with a 20 nm hole. (**d**) Optical force and light intensity vs illumination wavelength, at the opening of the nanohole. The optical force is assumed to be exerted on a gold nanosphere with the radius of $d = 15$ nm at $\lambda = 600$ nm. The arrows in blue color represent the optical gradient force. Reprinted from [28] under the Creative Commons Attribution License

determined by the nanohole diameter rather than by the incident wavelength. The proposed nanohole-structured mesoscale particle has an advantage that the spatial region of light confinement and enhancement can be tailored by choosing the geometry, 3D shape, orientation and dimensions of the nanohole [28].

5.3   Concept of Photonic Hook Scalpel

Today one of the key elements of the medical device is a laser scalpel connected with an optical fiber to add of flexibility [30]. The idea of application of a photonic hook effect as the ultraprecise laser scalpel for the first time was offered in [31]. In [32] we report on the formation of a curved optical hook created by fiber with mesoscale hemispherical particle with partially blocking fiber tip by an amplitude mask [33].

To illustrate the effect of the irradiating beamwidth on the PH formation, Fig. 5.6 shows the power flow patterns of the PHs formed by fiber with hemispherical particle tip with different mask. The width of the irradiating beam is regulated by the height of the amplitude mask. One can see that the curvature and the length of the PH change as the mask height increases. However, the PH position (point with maximum field intensity along PH) moves inside the fiber tip when the height of the amplitude mask is more than $0.35d$.

Simulations have shown that the spatial shape and curvature of the PH with FWHM below the diffraction limit [34, 35] can be simply tailored by varying the metal or absorption mask inserting into the fiber tip [36]. The correct combination of the illuminating beamwidth vs mask height enables the efficient beam shaping and directionality of the PH launching. Fiber-based photonic hook enables to optimize tissue cutting modes due to controllable curved shape of focusing area, taking into account that the orientation of this PH with a tissue contact strongly determines the heated area. In practical applications, a gel-like medium may be used to mimic optical properties of a biological tissue [37]. It could be noted, that in combination with optical tweezer laser scalpels can cut the particle under investigations.

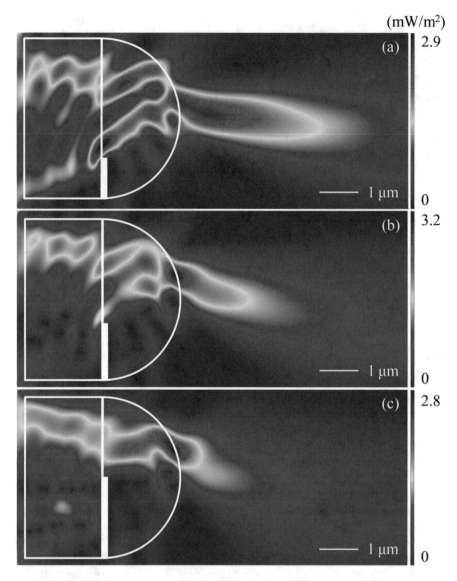

Fig. 5.6 Power flow patterns of the photonic hooks formed by fiber tip with metallic mask at (**a**) $h = 0.25d$, (**b**) $h = 0.35d$, and (**c**) $h = 0.5d$ (*h* is a mask height). Adapted from [32] with the permission from SPIE

5.4 A Hot Spot for Particle Trapping

Recently, authors of [38] reported on the generation of acoustic hook beams [39] using a holographic lens for particle manipulations [40]. It is interesting to note that compared with the conventional Airy-like curved beams, acoustic hook bends not from the starting point but at a certain position during beam propagation, which seems to disobey the conservation of momentum [14]. In [38], acoustic bottle beam creation was developed by rotating the lens that formed the hook beam and then trapping Rayleigh particles in this bottle beam in water. Note that in [38] the Janus particle was not a 3D structure, but a quasi-2D one, since its thickness (50 mm) exceeded its width (22 mm) by more than 2 times. Another approach to acoustic lens design was mentioned in [41].

It is of great interest to use focused surface acoustic waves in different area including biology [42], medicine [43], nondestructive testing [44], and microparticle manipulation [45–47]. For the hydroelastic waves, which propagate at the surface of water covered with an elastic sheet, it is also possible to observe an analog of an acoustojet. In [48], it is reported about the possibility of controlling hydroelastic waves up to the subwavelength scale. The main idea [48] is as follows. The elastic film surrounded by water, where both tension and flexural waves occur, is ubiquitous in nature [49] and illustrates the principal features of elastic surface waves. The general theoretical expression for the dispersion relation of hydroelastic waves was considered in [49, 50]. In the approximation of a plane wave, satisfying Laplace's equation, using the kinematic boundary condition on the sheet surface and assuming negligible sheet inertia and infinite depth, the linear dispersion relation, up to a constant factor, may be written in the form [9]:

$$\omega^2 = gk + \frac{Tk^3}{\rho} + \frac{Dk^5}{\rho} \tag{5.1}$$

where

$$D = Eh^3 / \left(12\left(1 - v^2\right)\right) \tag{5.2}$$

is the modulus of rigidity (or bending modulus) which depends sensitively on film thickness h (as well as Young's modulus E and Poisson's ratio v), ϱ is a density of material, T is a uniform and isotropic mechanical tension of a film, $\omega = 2\pi f$ is the pulsation, $k = 2\pi/\lambda$ is the wave number, and g is the acceleration of gravity.

The equation of the linear dispersion relation Eq. (5.1) involves three terms: a gravity one and two elastic ones. The third term of the right-hand side of Eq. (5.1) corresponds to bending, while the second one corresponds to capillary waves. For thicker films the flexural third term in Eq. (5.1) is leading so for hydroelastic waves the dispersion relation Eq. (5.1) must be rewritten as

$$\omega(k) \sim \left(\frac{D}{\rho} k^5\right)^{1/2} \tag{5.3}$$

From the equation Eq. (5.3), it is followed that the phase velocity, defined as $V = \omega/k$, depends on the D/ϱ ratio and the wave number k only. A relative effective refractive index n is $n \sim 1/V$. So taking into account the definition of flexural coefficient D Eq. (5.2), relative effective refractive index n can be achieved by changing three parameters: locally Young's modulus E, Poisson's ratio v, or the elastic film thickness h.

Let's consider two areas covered with elastic films (Fig. 5.7) with different thicknesses h_1 and h_2 and the same Young's modulus E and Poisson's ratio ν. In this case the ratio of their refractive indices n_1 and n_2 may be determined as follows. Taking into account that from Eq. (5.3) the wave number $k \sim \left(\dfrac{1}{D}\right)$ and from Eq. (5.2) $D \sim h^3$, we may write

$$n_1 / n_2 \sim V_2 / V_1 \sim k_2 / k_1 \sim \left(h_2 / h_1\right)^{3/5} \qquad (5.4)$$

It means that in the hydroelastic regime and considered approximation, the effective index contrast does not depend on wave vector k and is given only by the local film thickness h Eq. (5.4). Thus thinner film region corresponds to higher refractive indices and vice versa.

In [48] there is experimental observation of a FWHM of a focal spot of about 0.33λ. Since the thickness of the film and the effective refractive index of the structure are unambiguously related Eq. (5.4), this method is suitable for the formation of hydroelastic waves of an arbitrary structure by choosing the appropriate profile of the notch in the film.

5.5   Reflection Mode Photonic Hook

Recently, a specular-reflection (s-PNJ), which is a specific type of PNJ, originated from the constructive interference of direct and backward propagated optical waves focused by a dielectric microparticle located near a flat reflecting mirror [51, 52],

Fig. 5.7 Structure of the film in water surface

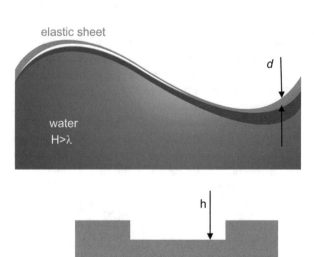

was discovered in [53, 54]. The unique property of s-PNJ is the possibility to break so-called fundamental "less than 2" [2] refractive index contrast limit for maintaining the spatial localization and intensity of PNJ when using dielectric particles with high refractive index contrast when regular photonic nanojet is not formed. The physical concept of an optical tweezer in reflection mode integrated into the microfluidic device was proposed [53, 54]. Importantly, such an optical trap shows twice as high stability to Brownian motion of the captured nano-bead as compared to the conventional PNJ-based traps. Preliminary simulation shows that the formation of photonic hooks in this mode is possible when oblique incidence of radiation or an oblique (rotated) mirror is used [55].

A concave semi-cylindrical mirror [56] with a dielectric coating for efficient generation of a reflecting photonic hook was studied in Ref. [57]. It was shown that the asymmetric vortexes of Poynting vectors cause the bending PH in reflection mode. However, this method [57] is not new and was actually considered in acoustics in Ref. [41]. It is pertinent to note that in the formation of localized beams in the "reflection" mode, it is possible to control the modulation of the standing wave of the photonic jet or the photonic hook [68].

All in all, the unique and fantastic qualities of subdiffraction limited structured beam waists, and subwavelength-scaled curvature radius of the PH showed important application prospects in the fields of imaging, nano-manipulation, biology and medicine, nonlinear optics, integrated systems, both for electromagnetic waves and for acoustic waves, and both for free space and for surface waves [10, 11, 13, 14, 24, 28, 31–33, 36, 38, 39, 52–55, 58–67].

References

1. A. Heifetz, S. Kong, A. Sahakian, A. Taflove, V. Backman, Photonic nanojets. // *J.* Comput. Theor. Nanosci., *6*, 1979–1992 (2009).
2. B. Luk'yanchuk, R. Paniagua-Domínguez, I. V. Minin, O.V. Minin, Z. Wang, Refractive index less than two: Photonic nanojets yesterday, today and tomorrow // Opt. Mater. Express, *7*, 1820–1847 (2017).
3. Y. Li, H.-B. Xin, H.-X. Lei, L.-L. Liu, Y.-Z. Li, Y. Zhang & B.-J. Li. Manipulation and detection of single nanoparticles and biomolecules by a photonic nanojet. // Light Sci. Appl 5, e16176 (2016)
4. Y. Li, H. Xin, X. Liu, Y. Zhang, H. Lei, and B. Li. Trapping and detection of nanoparticles and cells using a parallel photonic nanojet array // ACS Nano 10, 5800–5808 (2016).
5. X. Cui, D. Erni, & C. Hafner, Optical forces on metallic nanoparticles induced by a photonic nanojet // Opt. Express 16, 13560–13568 (2008).
6. F. Valdivia-Valero, M Nieto-Vesperinas, Optical forces on cylinders near subwavelength slits: effects of extraordinary transmission and excitation of Mie resonances // Opt. Express 20, 13368–13389 (2012).
7. F. Valdivia-Valero, M. Nieto-Vesperinas, Optical forces on cylinders near subwavelength slits illuminated by a photonic nanojet. // Opt. Commun. 294, 351–360 (2013).
8. A. Neves, Photonic nanojets in optical tweezers. // J. Quant. Spectrosc. Radiat. Transf. 162, 122–132 (2015).

9. H. Wang, X. Wu, D. Shen, Trapping and manipulating nanoparticles in photonic nanojets. // *Opt. Lett.* 41, 1652–1655 (2016).

10. I. V. Minin, O. V. Minin, V. Pacheco-Peña, M. Beruete, Subwavelength, standing-wave optical trap based on photonic jets. // Quantum Electron. 46, 555–557 (2016).

11. I. V. Minin and O. V. Minin, "Device for forming the optical trap in the form of the photon hook," Patent of Russia 161207 (27 October, 2015).

12. I. V. Minin, O. V. Minin, *Diffractive Optics and Nanophotonics: Resolution Below the Diffraction Limit*, Springer, Cham (2015).

13. A. Ang, A. Karabchevsky, I. V. Minin, O.V. Minin, S. Suchov, A. Shalin, 'Photonic Hook' based optomechanical nanoparticle manipulator // Sci Rep 8, 2029 (2018).

14. I. V. Minin and O. V. Minin. Recent Trends in Optical Manipulation Inspired by Mesoscale Photonics and Diffraction Optics // J of Biomedical Photonics & Eng 6(2), 020301 (2020)

15. R. Schley, I. Kaminer, E. Greenfield, E. R. Bekenstein, Y. Lumer & M. Segev. Loss-proof self-accelerating beams and their use in non-paraxial manipulation of particles' trajectories. // Nat. Commun. 5, 6189 (2014).

16. J. Baumgartl, M. Mazilu, K. Dholakia, Optically mediated particle clearing using Airy wave-packets. // Nat. Photonics 2, 675–678 (2008).

17. N. Efremidis, Z. Chen, M. Segev, D. N. Christodoulides, Airy beams and accelerating waves: An overview of recent advances // Optica, 6, 686–701 (2019).

18. D. Papazoglou, S. Suntsov, D. Abdollahpour, & S. Tzortzakis, Tunable intense Airy beams and tailored femtosecond laser filaments // Phys. Rev. A 81, 061807 (2010).

19. I.V. Minin, C.-Y. Liu, Y. E Geints, O. V. Minin. Recent advantages in Integrated Photonic Jet-Based Photonics.// Photonics 7(2), 41 (2020)

20. L. Novotny, B. Hecht, *Principles of Nano-Optics*. (Cambridge University Press, 2006).

21. G. V. Shuvalov, K. V. Generalov, V. M. Generalov, M. V. Kruchinina, E. S. Koptev, O. V. Minin, and I. V. Minin. Physical principles of development of the State standard of biological cell polarizability // *Russian Physics Journal,* 60(11), 1901–1904 (2018).

22. A.S. Ang, I.V. Minin, O. V. Minin, S.V. Sukhov, A. Shalin, A. Karabchevsky. "Low-contrast photonic hook manipulator for cellular differentiation." Proc. of the 9th Int. conf. on Metamaterials, Photonic crystals and Plasmonics (2018).

23. T.-H. Liu, W.-Y. Chiang, A. Usman, and H. Masuhara, Optical Trapping Dynamics of a Single Polystyrene Sphere: Continuous Wave versus Femtosecond Lasers // J. Phys. Chem. C 120(4), 2392–2399 (2016).

24. M. Spector, A. S. Ang, O. V. Minin, I. V. Minin, and A. Karabchevsky, Temperature mediated 'photonic hook' nanoparticle manipulator with pulsed illumination // Nanoscale Adv. 2(6), 2595–2601 (2020).

25. M. Spector, A. S. Ang, O. V. Minin, I. V. Minin, and A. Karabchevsky. Photonic hook formation in near-infrared with MXene Ti3C2 nanoparticles // Nanoscale Adv. 2, 5312-5318 (2020)

26. I.V. Minin, O. V. Minin, L. Yue, Z. Wang, V. Volkov, D. N. Christodoulides. Photonic hook – a new type of subwavelength self-bending structured light beams: a tutorial review // arXiv:1910.09543 (2019)

27. Y. Cao, Z. Liu, O. V. Minin, I. V. Minin, Deep Subwavelength-Scale Light Focusing and Confinement in Nanohole-Structured Mesoscale Dielectric Spheres // Nanomaterials 9(2), 186 (2019).

28. I. V. Minin, O. V. Minin, Y. Cao, Z. Liu, Y. Geints and A. Karabchevsky. Optical vacuum cleaner by optomechanical manipulation of nanoparticles using nanostructured mesoscale dielectric cuboid // Sci Rep 9, 12748 (2019)

29. I.V. Minin, O. V. Minin, Yu. E. Geints, E. K. Panina, and A. Karabchevsky. Optical Manipulation of Micro- and Nanoobjects Based on Structured Mesoscale Particles: a Brief Review // Atmospheric and Oceanic Optics 33(5), 464–469 (2020).

30. G. Keiser, F. Xiong, Y. Cui, and P. Shum, Review of diverse optical fibers used in biomedical research and clinical practice // Journal of Biomedical Optics 19(8), 080902 (2014)

31. O. V. Minin, I. V. Minin, N. Kharitoshin. Microcubes Aided Photonic Jet Scalpel Tips for Potential Use in Ultraprecise Laser Surgery. 2015 International Conference on Biomedical Engineering and Computational Technologies (SIBIRCON), 28–30 Oct. 2015, p.18–21.

32. I. V. Minin, O. V. Minin, Y.-Y. Liu, C.-Y. Liu. Concept of photonic hook scalpel generated by shaped fiber tip with asymmetric radiation // arXiv: 2007.11378 (2020).

33. I. V. Minin, O. V. Minin, C.-Y. Liu, and H.-D. Wei, Y.Geints, A. Karabchevsky. Experimental demonstration of tunable photonic hook by partially illuminated dielectric microcylinder // Optics Letters 45(17), 4899–4902 (2020)

34. E. Abbe, Beiträge zur Theorie des Mikroskops und der mikroskopischen Wahrnehmung // Archiv für Mikroskopische Anatomie, 9, pages 413–468 (1873); https://doi.org/10.1007/BF02956173

35. Lord Rayleigh F.R.S. XXXI. Investigations in optics, with special reference to the spectroscope // The London, Edinburgh, and Dublin Philosophical Magazine and Journal of Science, 8(49), 261-274 (1879); https://doi.org/10.1080/14786447908639684

36. Y.-Y. Liu, C.-Y. Liu, I. V. Minin, O. V. Minin. Subwavelength photonic hook generated by shaped fiber tip with asymmetric terahertz radiation. // Proc. SPIE 11582, Fourth International Conference on Terahertz and Microwave Radiation: Generation, Detection, and Applications, 15820E (17 November 2020); https://doi.org/10.1117/12.2579807

37. Tuchin, V. V., [*Tissue Optics: Light Scattering Methods and Instruments for Medical Diagnostics*, 3rd ed.], vol. PM 254, SPIE Press, Bellingham, WA (2015).

38. X. Ren, Q. Zhou, Z. Xu, and X. Liu. Acoustic hook beam lens for particle trapping // Applied Physics Express 13, 064003 (2020)

39. C. Rubio, D. Tarrazó-Serrano, O. V. Minin, A. Uris, I. V. Minin. Acoustical hooks: A new subwavelength self-bending beam // Results in Physics 16, 102921 (2020)

40. D. Sukhanov, I. V. Minin, O. V. Minin, I. Kuzmenko, T. Muksunov, E. Sivkov, F. Emelyanov. Control of levitating particle in ultrasound field // MATEC Web of Conferences 155, 01017 (2018)

41. Q. Zhou, J. Zhang, Z. Xu and X. Liu. Acoustic interference lens for trapping micro-scale particles // J. Phys. D: Appl. Phys. 52, 455302 (2019)

42. E. F. Gomez, M. Berggren, and D. T. Simon. Surface Acoustic Waves to Drive Plant Transpiration. // Scientific Reports, 7, 45864 (2017)

43. N. Sivanantha, C. Ma, D.J. Collins, M. Sesen, J. Brenker, R. L. Coppel, A. Neild, and T. Alan. Characterization of adhesive properties of red blood cells using surface acoustic wave induced flows for rapid diagnostics. // Applied Physics Letters 105, 103704 (2014)

44. Y. Ohara, T. Oshiumi, H. Nakajima, K. Yamanaka, X. Wu, T. Uchimoto, T. Takagi, T. Tsuji, and T. Mihara. Ultrasonic phased array with surface acoustic wave for imaging cracks // AIP Advances 7, 065214 (2017)

45. A. Ozcelik, J. Rufo, F. Guo, Y. Gu, P. Li, J. Lata and T.J. Huang. Acoustic tweezers for the life sciences. // Nature Methods 15, 1021–1028 (2018)

46. X. Qi, Q. Tang, P. Liu, I. V. Minin, O. V. Minin, J. Hu, Controlled concentration and transportation of nanoparticles at the interface between an ordinary substrate and droplet // Sensors & Actuators: B., 274, 381–392 (2018)

47. Q. Liu, J. Hu, I. V. Minin, O. V. Minin, High-performance ultrasonic tweezers for manipulations of motile and still single cells in a droplet // Ultrasound in Medicine and Biology, 45(11), 3018–3027 (2019)

48. L. Domino, M. Fermigier, E. Fort and A. Eddi. Dispersion-free control of hydroelastic waves down to sub-wavelength scale // EPL, 121, 14001 (2018)

49. J.-C. Ono-dit-Biot, M. Trejo, E. Loukiantcheko, M. Lauch, E. Raphael, K. Dalnoki-Veress, and T. Salez. Hydroelastic wake on a thin elastic sheet floating on water. // Phys. Rev. Fluids 4, 014808 (2019)

50. J. Bae, T. Ouchi, and R. C. Hayward. Measuring the Elastic Modulus of Thin Polymer Sheets by Elastocapillary Bending. // ACS Appl. Mater. Interfaces 7, 14734−14742 (2015).

51. I. V. Minin, C.-Y. Liu, Y.-C. Yang, K. Staliunas and O. V. Minin. Experimental observation of flat focusing mirror based on photonic jet effect // Sci Rep 10, 8459 (2020)

52. L. Yue, B. Yan, J. Monks, R. Dhama, Z. Wang, O. V. Minin, and I. V. Minin. Photonic Jet by a Near-Unity-Refractive-Index Sphere on a Dielectric Substrate with High Index Contrast // Ann. Phys. (Berlin) 1800032 (2018)

53. Y. E. Geints, A. Zemlyanov, I.V. Minin, O.V. Minin. Overcoming refractive index limit of mesoscale light focusing by means of specular-reflection photonic nanojet // Optics Letters, 45(14), 3885–3888 (2020)

54. I.V. Minin, Y. E. Geints, A. Zemlyanov, O.V. Minin. An extensive study of specular-reflection photonic nanojet: Physical basis and optical trapping application // Opt Express 28(15), 22690–22704 (2020)

55. Y.E. Geints, A.A. Zemlyanov, I.V. Minin, and O.V. Minin. Specular-reflection photonic hook generation under oblique illumination of a super-contrast dielectric microparticle // ArXiv: 2009.14012 (2020)

56. W. Yang, R. Gao, Y. Wang, S. Zhou, J. Zhang. Reflective photonic nanojets generated from cylindrical concave micro-mirrors // Applied Physics A 126, 717 (2020)

57. C.-Y. Liu, H.-J. Chung, and Hsuan-Pei, E. Reflective photonic hook achieved by a dielectric-coated concave hemicylindrical mirror // J. Opt. Soc. Am. B 37, 2528–2533 (2020)

58. I. V.Minin, O. V. Minin, G. Katyba, N. Chernomyrdin, V. Kurlov, K. I. Zaytsev, L. Yue, Z. Wang, and D. N. Christodoulides. Experimental observation of a photonic hook // Appl. Phys. Lett. 114, 031105 (2019)

59. O. V. Minin, I. V. Minin, K. I. Zaytsev, G. Katyba, V. Kurlov, L. Yue, Z. Wang, "Electromagnetic field localization behind a mesoscale dielectric particle with a broken symmetry: a photonic hook phenomenon," Proc. SPIE 11368, Photonics and Plasmonics at the Mesoscale, 1136807 (2 April 2020)

60. I.V. Minin, O.V. Minin, I. Nevedof, V. Pacheco-Peña, M. Beruete. Beam compressed system concept based on dielectric cluster of self-similar three-dimensional dielectric cuboids // An International Joint Conference of The 9th Global Symposium on Millimeter-Waves (GSMM 2016) and The 7th ESA Workshop on Millimetre-Wave Technology and Applications, June 6–8, 2016, Aalto University, Espoo 64–66

61. L. Yue, B. Yan, J. Monks, Z. Wang, N. T. Tung, V. D. Lam, O. V. Minin, and I. V. Minin. A millimetre-wave cuboid solid immersion lens with intensity-enhanced amplitude mask apodization. // Journal of Infrared, Millimeter, and Terahertz Waves, 39(6), 546–552 (2018)

62. G. M. Katyba, N. V. Chernomyrdin, I. N. Dolganova, A. A. Pronin, I. V. Minin, O. V. Minin, K. I. Zaytsev, and V. N. Kurlov. "Step-index sapphire fiber and its application in a terahertz near-field microscopy", Proc. SPIE 11164, Millimetre Wave and Terahertz Sensors and Technology XII, 111640G (18 October 2019)

63. N. Chernomyrdin, A. Kucheryavenko, G. Kolontaeva, G. Katyba, P. Karalkin, V. Parfenov, A. Gryadunova, N. Norkin, O. Smolyanskaya, O. V. Minin, I. V. Minin, V. Karasik, K. I. Zaytsev, "A potential of terahertz solid immersion microscopy for visualizing subwavelength – scale tissue spheroids," Proc. SPIE 10677, Unconventional Optical Imaging, 106771Y (24 May 2018)

64. I. V. Minin and O. V. Minin. "System of microwave radiovision of three-dimensional objects in real time", Proc. SPIE 4129, Subsurface Sensing Technologies and Applications II, (6 July 2000)

65. H.-H. N. Pham, S. Hisatake, O. V. Minin, T. Nagatsuma and I. V. Minin. Enhancement of Spatial Resolution of Terahertz Imaging Systems Based on Terajet Generation by Dielectric Cube // APL Photonics 2, 056106 (2017)

66. I.V.Minin and O.V.Minin. "Subwavelength self-bending structured light beams." In: Proc. of the Fourth Russian-Belarusian Workshop "Carbon nanostructures and their electromagnetic properties", Tomsk, Apr. 21–24, 2019. P.52–57.

67. S. Perez-Lopez, P. Candelas, J. M. Fuster, C. Rubio, O. V. Minin, and I. V. Minin, Liquid-liquid core-shell configurable mesoscale spherical acoustic lens with focusing below the wavelength // Appl. Phys. Express 12, 087001 (2019).

68. K. A. Sergeeva, A. A. Sergeev, O. V. Minin, I. V. Minin. A closer look to photonic nanojet in reflection mode: control of standing wave modulation //Photonics 8(2), 54 (2021).

Chapter 6
Conclusions

UNESCO declared 2015 to be the International Year of Light and Light-based Technologies, and the twenty-first century is known now as the "century of the photon." Humanity is trying to replace electronic devices with photonic ones. However, today the methods of manipulating and controlling photons at the micro- and nano-levels are still more scientific than technological. At the same time, a certain qualitative progress in this area can be achieved using the phenomena of interference and diffraction at the mesoscale. This will transform the light from its original properties in the radiation source to some set of properties specified by the developer. To miniaturize optical systems, it is necessary to take into account the near-field effects at distances around the wavelength of light. One such example is the concept of photonic nanojets, when light interacts with mesoscale symmetric objects and focuses on a subwavelength scale. The nature of photonic nanojets is that the relative lower refractive index dielectrics have relative low-quality factor modes. On the other hand, they can both spatially and spectrally overlap with each other due to the momentum of light being small in low refractive index media. The interference of these modes can form a localized field, known as photonic jet effect. Photonic jets were initially suggested for the realization of subwavelength field localization as a lens, and later they became a paradigm for structuring and controlling propagation of waves, for example, a photonic hook concept. The study of the interaction of light with asymmetric objects (Janus particle – a class of two-faced 3D mesoscale dielectric particles), the degree of asymmetry of which can be considered as an additional degree of freedom for controlling the incident light, led to the discovery of the photonic hook effect for the first time in 2015. Moreover, for photonic jet, the refractive index contrast has to be less than two (in geometrical optics approximation), so only one degree of freedom is to increase the size of the particles to increase the level of field localization beyond the diffraction limit (if we do not consider the introduction of additional elements into the system, e.g., masks for the implementation of the effect of anomalous apodization). But in the case of specular reflection, this limit may be breaking and the degree of freedom increases.

O. V. Minin, I. V. Minin, *The Photonic Hook*, SpringerBriefs in Physics,
https://doi.org/10.1007/978-3-030-66945-4_6

Surprisingly, despite PHs having a curved configuration, their properties are more inclined to the normal light in our common sense. Photonic hooks are unique, work on a scale much smaller than Airy beams, and fundamentally differ from Airy-family beams, as their radius of curvature is several times smaller than their wavelength. This is the first time that such a small curvature radius of electromagnetic, acoustic, and plasmonic waves has been recorded. In contrast to Airy beams, which are usually initialized by a bulky laser paired with an expensive spatial light modulator, the photonic hook can be generated using only wavelength-scaled particles and a light source. One idiom supported by people since ancient times is "seeing is believing." These unconventional electromagnetic, acoustic, and surface structured beams not only have been realized in experiment but also have wide application potentials at the micro- and nano-scale, such as for particle manipulation, super-resolution imaging, surface plasmons, surface modification, optical switches, frequency-division demultiplexer, on-chip bendable light communications, waveguides (including curvilinear), etc. *One of the interesting areas of the future research may be the application of Janus particles in terahertz photoconductive antennas. The shape of the terahertz photoconductive antenna substrates can be changed from a classic rectangle to an arbitrary three-dimensional shape, including the Janus particle, to improve its directivity and ef-ficiency of the entire THz antenna without adding other structures or conventional lenses.*

The first of such simple structured field was constructed from combinations of dielectric mesoscale cube diffraction with prism refraction, and today the Janus particles with asymmetric external shape, asymmetric internal structure, or symmetric particle with illumination with broken symmetry – the subwavelength localization of radiation in curved space – can be controlled and manipulated in unprecedented ways. Such symmetry breaking can be used for the further improvement of the properties of the localized beams. Full-wave simulations promise to uncover hidden functional relationships between the physical properties of the near-field structured beams that we can observe. By controlling beam curvature by field polarization, illuminating wavelength or illuminated field structurization, the technique could accommodate different wave-scaled applications, making them possible to use in highly integrated circuits due to exhibition of subwavelength field localization. For example, in our research it was experimentally shown for the first time as in optics as in terahertz that when switching two orthogonal states of linear polarization of the incident wave, the photonic hook changes the direction of its curvature to the opposite. It is noteworthy that the generation of photonic hooks is possible, both insensitive and sensitive to polarization. Importantly, such a mechanism can be readily translated to other non-optical waveforms such as acoustic, plasmonic, surface, and electronic waves. Moreover, the dynamic formation of Janus particles allows both control of the shape of the radiation localization region and opens the new way for obtaining new functional properties of structured fields. Similar effects can be observed in atmospheric physics, dew in nature, etc. For example, when a water drop freezes, its phase state changes from liquid to solid (ice). These materials have different properties and, in particular, different refractive index, which allow to dynamically form a photonic hook.

Note that the creation of unusual structured localized fields is possible not only outside the dielectric object but also inside it. The figure shows the structure of the field inside a cylinder consisting of three materials with different optical contrast.

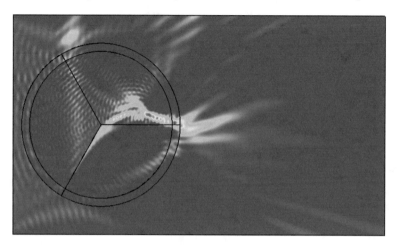

An example of the structure of a field inside a cylinder consisting of three materials with different optical contrast. In such materials, it is possible to implement a meso-object, which, when irradiated from one side, forms a local localized region such as a photonic jet, and when irradiated from the opposite direction, for example, photon claws (twin photonic hooks)

This short book not only reveals the explicit physical role of any given types of photonic hooks in optics, plasmonics, and acoustics, but to our point also provides an alternative design roadmap of a new subwavelength curved structured light. This area of photonics, optics and acoustics is developing rapidly, and this book is just only a brief introduction to it. We hope this book allow the unlocking of novel functionalities of this new class of subwavelength self-bending beams and bring about disruptive performance improvements of future high-integrated circuits and enable realization of various chip-level photonic devices.

Index

© The Author(s), under exclusive license to Springer Nature Switzerland AG 2021   87
O. V. Minin, I. V. Minin, *The Photonic Hook*, SpringerBriefs in Physics,
https://doi.org/10.1007/978-3-030-66945-4

Printed in the United States
by Baker & Taylor Publisher Services